The Righteous Way

GOLDEN JUBILEE EDITION

Starmel Allah

The Righteous Way: Golden Jubilee Edition

ISBN 978-0-692-55006-9

Visit us on the web at
www.therighteouswaybook.com

Dedication

To the one who made us equal to himself—Allah, the visionary and the founder of this culture and nation. To the mighty Five Percent Nation of Gods & Earths or "NGE." To the Smith and Jowers families. To the everlasting memory of the Messenger, Hon. Elijah Muhammad and his family, to Sister Dora 3X Smith; To the everlasting memories of Justice, Rahim, Malik, Latif, Jahad and Justice (of "The Harlem Six"), Black Messiah, BismiAllah, Prince Allah, Uhuru, Kihiem, Al-Jamel, Akbar, Al-Salaam, Wise Jamel, Dumar Wa'de Allah, Dr. Janal Allah, and all their families. To the memory of former mayor of New York City, John V. Lindsay and his aid, the Mayor's Man, Barry H. Gottehrer. To the everlasting memories of Universal Shaamgaudd Allah, Byheen, Siheem, Sha Sha, Bali, Gamal, Hasheem, Lakee, Ali, Ahmad, Allah-U-Akbar, Khrist, Allah Calib Allah, Islam Ubeeka, Lord Reveal, Allah Mathematics, God B, Prince I Allah (South Jamaica, Queens), Black God, Akmin, Harmeen, Amar Self (Power Hill), Azreal, Dr.Saviour Shabar U Allah, Magnetic Allah, Sha Mel Allah, Pure Life Allah, Universal Zig Zag Zig Allah, Lord Justice, Positive Ramel Allah and all their families. To Sister Carmina, Makeeba,

Yokay, Saleema, Queen Shavasia Earth, Queen Qualisha Earth, Mecca Wise; Sister Maryann, and to the many Gods and Earths who made it possible in ways big and small for making this a great culture, but whose names couldn't possibly fit within the borders of these pages. May peace be upon them and may they Rest in Perfection. To Abu Shahid the Elder, the Father's Brown Seed; Eye-God; ABG; Gykee Mathematics; Akim; Raleak; Knowledge God; God Supreme; Allah Born (Allah B); Um Allah; Lord Dumar Allah; C-Allah (Head of Medina); Queen Omala; Niheem; Demina; Omina; El-Latesha; Mecca; Gevasia; Kareema; Queen Cipher. To righteous people, and those striving to be righteous. To the poor, disenfranchised, uneducated, underprivileged, unemployed and marginalized. To those seizing the time and claiming their destiny. To Hip Hop's legends. To the modern day saviors, savants, gurus, and sages born out of struggle. To the keepers of the flame. To those fighting for freedom, justice and equality. To those who right wrongs. To teachers and tutors. To all the youth who have yet to find their identity, voice and strength. To those who teach the righteous way, those who have lost their way, and those who have yet to find their way. To the whole Rosedale. To my brothers and sisters throughout Queens, Brooklyn, Manhattan, the Bronx, and Staten Island. To Long Island. To all students. To my fellow students, staff and professors at Queensborough Community College. To my family and friends at home, all over the country, and around the world.

Contents

FOREWORD

Shahid M. Allah

THE CIRCUIT OF the base of our Great Pyramid (of Khufu in the Land of the Black Ones/Kemet) divided by twice its height produces the numerical value (3.14) for: Π Pi), the 16th symbol or letter in the so-called Greek Alphabet. The 16th symbol or letter in the English Alphabet is P. In the Supreme Alphabet of the GODS and EARTHS the 16th letter stands for POWER! Decades ago when Allah presented the Jewels of the Nation of Islam's Problem No. 13 out of the PROBLEM BOOK to His young *Five Percent (5%)* many among the GODS and EARTHS, and on the outer perimeter (in society, in general) were not aware of the **significant degree of power** that Allah had bequeathed unto His students. Those *ten numbers and twenty-six letters* were/ are a very sacred, valuable tool similar to the Masons Geometry and the learned, elders (of the Ancient Oracles of Wisdom and Light in East Asia, otherwise known as *Africa*) who utilized the mathematical science of **gematria*** to codify their various scientific experiments, formulations and technology. We live in a time wherein although the learned among us know and understand the usage of such science it, along with other higher degrees of knowledge, to a large extent, are still *mystified* by TV programs such as the Discovery Channel, et cetera, who would rather give *credit* to our Inter-Galactic Star Brethren (whom they choose to call Aliens) for the design and creation of the pyramids, ziggurats, sphinx, Grand Lodge of Luxor, et cetera, instead of bearing witness that the Original, Black Man of Asia (Planet Earth) is the Master Architect Who designed and constructed these wonders of the world!

"In the 1st Dynasty of Kemet the King donned a red fez for a crown with an emblem of a bee or hornet on that fez; and he ruled from a palace known as the Red House. The red was for the Sun; and the bee was symbolic of the code of work of raising food like a bee does honey. The bee stinger emblem was a sign of this king's mighty warriors. This particular King and his people existed in what is known as the Delta region of lower Kemet (Egypt)." † Interesting The Bee is the title of the **16th** Sura (Chapter) of the Holy Qur'an. And, the 16th degree of Lost Found Muslim Lesson No. 2 of the Nation of Islam reads as follows: "16. Who is the 5% of the Poor Part of the Planet Earth? ANS.- They are the poor, righteous Teachers, who do not believe in the teachings of the 10%; and are all-wise; and know who the Living God is; and Teach that the Living God is the Son of Man, the Supreme

* Gematria is the calculation of the numerical equivalence of letters, words, or phrases, and, on that basis, gaining, insight into interrelation of different concepts and exploring the interrelationship between words and ideas.

† Excerpt from © 2015 revised edition of THY KINGDOM COME: Dispelling the Greek Myth in the Black Fraternal System by Shahid M. Allah.

Being, the (Black Man) of Asia; And Teach Freedom, Justice and Equality to all the human family of the Planet Earth. Otherwise known as: Civilized People. Also are: Muslim and Muslim Sons."

October 10th, 2014 on the Gregorian calendar marked the fiftieth (50th) Anniversary of the Nation of GODS and EARTHS. The number fifty (50) is denoted, in *5%* terms, by a Power [5] and a Cipher [0]. Starmel Allah, a rising literary *star* amidst the GODS and EARTHS, *hailing* like *sleet* from the Desert[*], presents to you in this book: Volume II of The Righteous Way- The NoGE[†] 50th Anniversary Jubilee Edition! This event was indeed an overwhelming, extraordinary one; because it was/is a celebration of a genuine generation of GODS (& Goddesses) who, against the odds pitted against them by the F.B.I.'s engineered counter-intelligence program ("cointelpro") that put drugs and other *goblins* in our urban areas, this supreme generation **rose** up from the grave of ignorance and decadence to shine like diamonds!

Fifty (50) years ago Allah bequeathed The Supreme Treasure unto the Babies (Youth) of *Mecca* [‡]and the New York City tri-state area in the pure form of Supreme Mathematics, Supreme Alphabet[§] and One Hundred and Twenty Lessons. And, with these *Keys* Allah commenced to remove the *rusty locks* that had our babies bound (for a one-way trip to hell)! By way of these life-giving teaching, whose inception started with Master Fard Muhammad raising up The Honorable Elijah Muhammad, the GODS and EARTHS

* The borough of Queens, New York, in 5% terminology is referred to as the "Desert".

† "NoGE" is an acronym that stands for "Nation of GODS and EARTHS.

‡ "Mecca" is the 5% term used to describe the borough of Manhattan, New, and Harlem, in particular.

§ See Nation of Islam Problem Book, Problem No. 13, as it was given to the Honorable Elijah Muhammad By Allah in the Person of Master Fard Muhammad.

have produced a culture that *runs circles (circuits) around* Satan and his limited way of life! Allah Manifested True and Living Super Heroes (and Heroines) for the whole wide world to see!

In this fantastically stimulating and motivational book you will get a real good glimpse of just what the big celebration was about back on the weekend of October 3–5, 2014, when GODS and EARTHS from various fields of endeavor, and different parts of the nation and globe convened in *Mecca* to celebrate the 50 NoGE! From the loins of the GODS and EARTHS have come not only some of very best who have ever done it in entertainment, but many others who have and continue their *shine on* in the realm of law, medicine, engineering, sociology, psychology, medicine, aerospace, culinary arts, construction management, clothing, sales, martial arts, education, military, politics, clairvoyance, et cetera!

Starmel Allah, in my humble opinion, is definitely one of those who has adhered to the Wise Words of the God Universal Shamgaudd Allah (May the Peace ad Blessings of Allah Be Upon Him) when he once wrote: *"If you are going to publish the Truth kick it all the way live, right and exact! Don't make our job out here on the streets any harder than it already is; and don't be afraid to ask the First Borns and Elders for guidance when you need it."* Universal Shamgaudd reminded us on how Allah once admonished him with regards to his (then, back in the 1960s) ideas of how he would utilize his powers.

On October 7, 1985, amidst his P.O.W.E.R., At Last! Forever! Phenomenal lecture given at Madison Square Garden in midtown Manhattan, NY, the Honorable Minister Louis Farrakhan specifically addressed his *Five Percent brothers of the Five Percent Nation,* and has always held the GODS and EARTHS in high regard! We now stand in a pivotal TIME wherein, By the Grace of Allah, the Honorable Minister Louis Farrakhan, with the help of many, including the GODS and EARTHS, has rebuilt the Nation of Islam; and the foundation of the Nation of GODS and EARTHS grows stronger and stronger on a daily basis. There is a wide, open field of growth and development out there for the wide awake, balanced man, woman and child!

So, just don't browse through this book, looking to see if your name is in the shout outs and/or your picture is on one or more pages. Read and study this book, so that you can draw up the very best part from it. Always remember that, as Righteous people, we are to preserve the best part for ourselves.

Writer, orator and public servant, Shahid M. Allah, is from Thompkins Housing Projects in Bedford-Stuyvesant, Brooklyn. He is notably the first published author to arise from Allah's Five Percent Nation of Gods & Earths. He is the co-founder of THE WORD newspaper (1987). He has written for: The Final Call newspaper, The National Newport News & Commentator, The City Sun, Your Black Books Guide and the Associated Press. His Self-published books include: Thy Kingdom Come (1989), Take A Second Look (1992), The Plot To Kill Clarence 13X (1995), and The Farrakhan Code (2000). Shahid has lectured at Syracuse University, New York University, Morgan State University, Old Westbury College, St. Johns University, Howard University, University of Mass., Immaculate Conception Seminary in Douglaston (NY), University of Maryland, University of Baltimore, various Muhammad Mosques in the USA, York College, Hunter College and University of Buffalo. He holds a B.S. Degree in Speech Communications & Marketing, Syracuse University, May 1986.

INTRODUCTION

In Perspective

WE ARE HALF way through a century of studying, discussing, dialoging, arguing (for some), and living out the foundational teachings of Allah. While some have given their very lives in dedication to this culture since our year one, some have just began to take their first steps on this journey of self-actualization, community empowerment and universal consciousness. Supreme Mathematics and Supreme Alphabets were given as keys to understanding the knowledge and wisdom we call 120 Lessons. The thirst for knowledge caused some to use 120 as the foundational blueprint to our learning, while others used it to augment their personal rhetoric and prejudices. Some became scholars and experts of 120 Lessons, while others became detractors, settling on a partial understanding of its context. There are students who empowered themselves and others who've dropped out of this school of divine thought to become Five Percent flunkies. As decades went by, teenagers with knowledge of self started having families and with that, increased responsibilities. Those who found the balance between work and maintaining a family continued teaching their children who became the new generation of Gods and Earths. Moreover, each Five Percent was required to teach 9 students which would multiply to the explosive growth and development of a nation within a nation. Each year, however, the Five Percent have been faced with challenges to their history, leadership, and legitimacy. Each time, few of the qualified were able to rise and meet these challenges and become morally victorious in the face of opposition.

The 50th (Golden Jubilee) Anniversary is a significant point in time for the Five Percent Nation of Gods & Earths. As you will come to learn, we were not supposed to be here according to the plans of those who oppose righteousness. But Allah is the best of planners. He chose this name and made us equal to himself through righteous teachings. He took young people to task fifty-one years ago as he taught them the meaning of civilization, righteousness, the knowledge of himself, and the science of everything in life. By his example, He showed them love, peace and happiness. A little over a half century, many have come and many have gone. Some have gone astray from this way of life and some have returned to the essence. Many who have departed have given their opinions and critizisms from the sidelines, while some of us have rolled up our sleeves to just do the work. Those who are here today are those who *"want"* to be here. They are those who *"chose"* to be here. Generations from now, people will be able to look back at how far we've come and see the proof of our growth and development. This edition looks beyond the historical context of the lessons and calls for leadership in the Five Percent duty to teach what we know by example. This, of course, won't be successful without personal and national change. This subject has proven to be too weighty to have simply been discussed on social media, therefore, a follow-up to the first edition of *The Righteous Way* was necessary to advance our mission of bringing light to the blind.

Allah's teachings were revolutionary in that they came at a time when the public only knew about Black Muslims in the Nation of Islam. He came out of the Nation of Islam's Temple #7 in Harlem, New York where the Messenger, the Honorable Elijah Muhammad's teachings were taught under the leadership of Malcolm X (El Hajj Malik El Shabazz). The teachings at the time, were providing Black people, and Black men in particular, with a message. The message was preparing *the way* for God to make himself known as described in the book of Malachi chapter 3, verse 1. His understanding of the knowledge brought by Master Fard Muhammad through his Messenger, Hon. Elijah Muhammad was made clear to his Five Percenters 34 years after 1930. Between 1964 and 1969, Allah taught his Five Percenters much about the righteous way. After June 13, 1969, they proved their word was bonded to the teachings as they emerged on their own as Gods & Earths. The nation is comprised of a culturally diverse group of people from all walks of life. We aren't bound by emotion. Rather, we are bound together by things that transcend emotions, by principles and laws, by ideals and truths found within the reasoning of science and mathematics. It's important to keep these points in mind as you read through this book and the first edition. We are also bound by the integrity of the Living Word made plain to show and prove consistency of truth as a living reality.

If you are like most Americans and indeed like multitudes of people across the globe—the media are a fixture in your daily life.

You may wake up to a radio show each morning or read a newspaper over breakfast or while commuting on a bus or train. You may watch televised news stories concerning the latest international or local crisis in the evening. You may receive real-time news updates on your cell phone, Internet news sites and blogs, or tune in to the 24-hour news channels available virtually everywhere. You may have also come across interviews, articles, essays and programs regarding this culture which may or may not have been accurate. There is no escaping that, as a citizen of the twenty-first century, you are inundated in a sea of news and information on a daily basis. Some of what the media offer you is meant to entertain, some is meant to inform, but increasingly the lines between the two have blurred. Somewhere within those blurred lines, the truth about this culture was skewed. We survived prejudice on many levels and each time we were victorious because we told the story according to how we lived it — the righteous way. Many have written about our past, however, this book will look into that past in order to envision the possibilities of our future.

CHAPTER 1

Now O'clock

WHY ARE THESE teachings relevant now? The conscious influence of the Five Percent culture on Hip-Hop has made it the most lucrative and impressionable cultures of our day. But the elements of knowledge and enlightenment are absent, with the exception of a few artists. Today's generation is living in a transitional moment called now, and now is the time for this generation to grow in their knowledge.

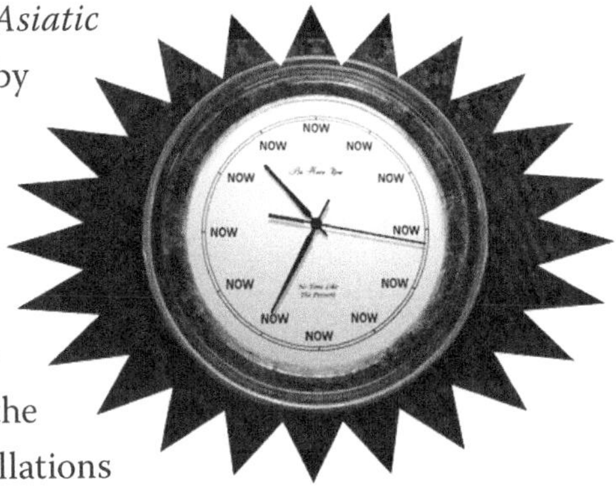

Time is a measure of motion. Depending on how much research we have done on the "*Asiatic Calendar*," as it is referred to by Elijah Muhammad, we'll see that it relates directly to our Earth's Precessional Cycle and records history in astrological periods of approximately 2,000 years. This is the great *clock* of the universe — the Zodiac (Constellations our planet and Sun cycles around approximately every 25,920 years). Each age (or sign is a group of stars whose effect on the planet is based on the planet's position) represents more or less 30 degrees of the cycle and 2,160 years of time. During this 2,160-span of time, the Zodiacal Sign identified by its position at the point of the vernal equinox is deemed the ruling sign under which people on Earth are affected philosophically, socially, and politically for the indicated period.

Many have spoken about the Age of Aquarius, but what is it? The Age of Aquarius is an astrological term denoting either the current or forthcoming astrological age, depending on the method of calculation. Aquarius comes from the Latin *aquarius,* literally *"water carrier,"* an adjective, *"pertaining to water."* Water is likened unto wisdom (enlightenment, judgment, truth, and reason). Astrologers maintain that an astrological age is a product of the earth's slow precessional rotation and lasts for 2,160 years, on average (1 degree every 72 years. 12 zodiac signs, 12 x 30 degrees each sign = 360 degrees. 30 degrees x 72 years = 2,160 years = 1 great month. 12 zodiac signs x 2,160 years per sign = 25,920 years). This correlates to the approximation given in our lesson that refers to the Original man renewing history every 25,000 years in relation to the Earth's circumference.* In a book entitled *The Biggest Lie Ever Told,* Malik H. Jabbar explains these astrological implications and provides approximate calculations for the Ages as follows:

> *PRECESSION MOVES WESTWARD THROUGH THE ZODIAC, THAT IS FROM THE THIRTIETH DEGREE OF A SIGN BACKWARD. THE BEGINNING OF THE 25,920 YEAR CYCLE RECOMMENCED 15,079 YEARS AGO (As of 1993 and assuming the 30th degree of Virgo as the commencement*

* 1st degree in the 1-40

of the Great Year cycle). THE NEXT CYCLE OF AQUARIUS WILL BEGIN IN THE YEAR 2034 AD. A CHART OF THE CYCLES IS AS FOLLOWS:

PRECESSIONAL AGES:

VIRGO COMMENCED 13086 BC
LEO COMMENCED 10926 BC
CANCER COMMENCED 8766 BC
GEMINI COMMENCED 6606 BC
TAURUS COMMENCED 4446 BC
ARIES COMMENCED 2286 BC
PISCES COMMENCED 126 BC
ACQUARIUS WILL COMMENCE 2034 AD
CAPRICORN WILL COMMENCE 4194 AD
SAGITTARIUS WILL COMMENCE 6354 AD
SCORPIO WILL COMMENCE 8514 AD
LIBRA WILL COMMENCE 10674 AD
AND VIRGO WILL REPEAT THE YEAR ONE 12834 AD[*]

According to Jabbar's calculations, the Age of Aquarius will begin on or about 2034.[†] The advent of the Five Percent in the west, the fall of governments, the decline of the American dollar, the countless people turning away from the lies in religion all indicate a turning point of events that will bring about the Age of Aquarius (or ***The Age of Enlightenment, Truth and Reason***). Perhaps we have already seen ourselves living in the time of enlightenment, truth and reason where people are questioning more, trying to make sense of things, and applying logic to establish and verify facts. Even those that claimed to have been enlightened, have truth and sound reasoning are awakened[‡] by those who are really enlightened, have truth and better reasoning.

---

* Malik H. Jabbar, *The Biggest Lie Ever Told*, Revised Edition, p. 3, footnote #2, Rare Books Distributors, Columbus, OH, 2002

† Just 20 years from 2014

‡ We called it *"getting sparked"*

Many astrologers believe that the Age of Aquarius has arrived recently or will arrive in the near future. On the other hand, some believe that the Age of Aquarius arrived up to five centuries ago, or will not start until six centuries from now. Despite all references provided by various sources, astrologers cannot agree upon exact dates for the beginning or ending of the ages (or their civilization). What we can say for sure is that a unique set of events concerning humanity has taken place in the last 500 years that have produced people who have had a significant influence on or will directly usher in the *Age of Aquarius* (a period of enlightenment and reason).

The Age of Aquarius predicted to start 2034 AD/ The Advent of Allah's teachings 1964 AD. Are we ahead of our time by 70 years? Or perhaps, we are right on time. Although no one may know the precise time of the new age, it is understood by those of us who are Five Percenters that God has come in his own good time. Empiricists and rationalists (inside and outside of the nation) may wrangle over the applicability of this subject, but several factors can't be denied.

Knowledge of Self has reawakened many in the West in a relatively small amount of time. True knowledge entered the minds of unenlightened people through Noble Drew Ali and Marcus garvey who learned from Dusé Mohamed Ali while in London. This occurred during the world's first global conflict, the Great War (also known as the *"European War"* or *"World War I"*) pitted the Central Powers of Germany, Austria-Hungary and the Ottoman Empire against the Allied forces of Great Britain, the United States, France, Russia, Italy and Japan. Meanwhile, enlightenment for us reached a pivotal point between 1930-1934 through Master Fard Muhammad and Elijah Muhammad. The intensity of self-awareness peaked during the Black Liberation and Civil Rights Movements. In this time, Allah revealed himself to the young and gave them the keys to life after 6,000 years of barbarism, religio-political imperialism, slavery and world conflict. Moreover, various events throughout history have shaped our reality today:

1914 – World War I ensued in Europe; The Clock of Destiny ticks as Noble Drew Ali establishes the Moorish Science Temple of America.

1934 – After Fard's disappearance, Elijah Muhammad emerged as the leader of the Nation of Islam; President Franklin D. Roosevelt establishes the New Deal.

1954 – The U.S. Supreme Court ruled segregated schools are unconstitutional in Brown v. Board of Education.

1974 – The Watergate Scandal exposed government corruption to the American people; Hip-Hop, largely influenced by the culture of the Gods & Earths, emerges.

1994 – The most revolutionary Hip-Hop group (Wu-tang Clan) lead by the Gods, changed the record industry and taught Allah's teachings to the world.

2014 - Gods & Earths celebrate the 50th Anniversary of the culture at the world famous Apollo Theater in Harlem, New York.

We are living in the 50th-year point of teaching our people the Knowledge of Self, righteousness and the meaning of being civilized. Some have among us studying from 35-50 years. And some have publically repudiated the lessons and Allah's teachings. So many have wasted an exorbitant amount of time and precious prana wrangling with people from other nations, cultures and movements. So many have argued and fought over these teachings and others are confused about them. People are questioning their faiths and governments have lost credibility. For all these reasons, an age or period of enlightenment, truth and reason in the world is due.

Today, the right to freedom of conscience threatens to unshackle the minds of the blind. Within our Supreme Mathematics, we find the correlation between the Sun, Moon and Stars and Man, Woman and Child. Astrology and astronomy were one subject when the Earth's first (Original and aboriginal) people made conscious attempts to measure, record, and predict seasonal changes by reference to astronomical cycles.

By 3,000 BC, scientists have found evidence that widespread civilizations had developed sophisticated awareness of celestial cycles, and are believed to have consciously oriented their temples to create alignment with the heliacal (appearance of stars over the eastern horizon) risings of the stars. These are things we forgot how to understand. Non-original people became students of this science but began to see that their time and power on Earth was limited through this study. So, up until about the 17th century A.D., this was no longer regarded as a science and became mixed, diluted and tampered with. English astronomers tried to make this work for their benefit but instead relegated it to an art. The mis-education of many in the west has caused our own to disregard it as well. Historian of science, Ann Geneva, put it this way:

> *"Astrology in seventeenth century England was not a science. It was not a Religion. It was not magic. Nor was it astronomy, mathematics, puritanism, neo Platism, psychology, meteorology, alchemy or witchcraft. It used some of these as tools; it held tenets in common with others; and some people were adept at several of these skills. But in the final analysis it was only itself: a unique divinatory and prognostic art embodying centuries of accreted methodology and tradition."* *

In popular U.S. culture, the Age of Aquarius refers to the advent of the New Age movement in the 1960s and 1970s. Allah began teaching others his realized identity circa late September/ early October 1964. During this time, the Earth's position would be in the sign of *Libra*, the seventh astrological sign in the Zodiac representing Justice. It is noteworthy to mention that Allah's right hand man's name was Justice and the chorus of the Five Percent national anthem goes *"Peace Allah, Allah and Justice."*

The true New Age Movement is one that teaches students to look inwardly as part of gaining the knowledge of their selves for self-improvement. The term *New Age* refers to the coming astrological Age of Aquarius. Astrologers believe humanity is affected by changes in astrological ages, namely, the rise and fall of civilizations, nations and superstitious beliefs. The world's reliance on technology, computers, information and humanitarian views are signs that we are upon that time.

This paradigm shift in mindset shapes a total self-awareness without borders or confining religious dogmas. It is inclusive

---

* Ann Geneva, *Astrology and the Seventeenth Century Mind: William Lilly and the Language of the Stars*, p.9.; Manchester University Press ND, 1995

and pluralistic. It holds to an altruistic and holistic worldview, emphasizing that the mind and body (God and the people) are not only interrelated, but are one reality. Man, woman, and child are the microcosm of the macrocosm we know as universe. It is a worldview shaped by science, reason and interrelationships between man and woman within the universe. Opposite of what world religions taught, Allah (God) was not invisible, not a formless spirit and he did not come out of the sky. Today, the greatest threat is for a Black man to call himself Allah (or God) and teach righteousness openly to others.

The Nation of Islam teaches that Allah came in the person of Master Fard Muhammad who came to Detroit to a place called "Black Bottom," where the people lived in the deep ghetto. He started teaching there, going door-to-door like an ordinary salesman. He revealed himself to and taught Elijah Muhammad who founded the Nation of Islam. As the teachings spread, others would soon learn that the God they have been searching for all these years existed right among them. A deeper meaning of the teachings came to the realization of one whom, with the knowledge of his self, took the name Allah and taught the young in New York City. The wise among us have maintained that Master Fard Muhammad brought the *Knowledge,* Elijah Muhammad brought the *Wisdom,* and Allah brought us the *Understanding.*

The identity of God was being unveiled through the Supreme Wisdom of a Black man making his self known. His identity would no longer be shrouded in mystery by secret societies in this new day and age. Mystics and esoterics such as Emanuel Swedenborg, Franz Mesmer, Eliphas Levi, Helena Blavatsky, and George Gurdjieff held some of the ancient philosophical principles that can be traced to the pre-existing wisdom written by the wisest of ancient Black civilizations. In essence, the so-called New Age movement is but a name given to the revival of fallen man and woman as a result of the work of enlightened individuals. However, the New Age is really found in the teachings of the Five Percent who know the truth and teach the truth to the people now during this period (or Age) of enlightenment.

This change did not go unnoticed by the highest intelligence officials in the country. In fact, it is widely known that J. Edgar Hoover, as Director of the FBI during this time, was responsible for the counterintelligence program that disrupted and destroyed the Black leadership during the Civil Rights era. The stated purpose of the COINTELPRO is as follows:

> *The purpose of this new counterintelligence endeavor is to expose, disrupt, misdirect, discredit, or OTHERWISE NEUTRALIZE [emphasis added] the activities of black nationalist hate-type organizations and groupings, their leadership, spokesmen, membership, and supporters, and to counter their propensity for violence and civil disorder. The activities of all such groups of intelligence interest to the Bureau must be followed on a continuous basis so we will be in a position to promptly take advantage of all opportunities for counterintelligence and inspire action in instanceswhere circumstances warrant. The pernicious background of such groups, their duplicity, and devious maneuvers must be exposed to public scrutiny where such publicity will have a neutralizing effect. Efforts of the various groups to consolidate their forces or to recruit new or youthful adherents must be frustrated. NO OPPORTUNITY SHOULD BE MISSED TO EXPLOIT THROUGH COUNTERINTELLIGENCE TECHNIQUES THE ORGANIZATIONAL AND PERSONAL CONFLICTS OF THE LEADERSHIPS OF THE GROUPS AND WHERE POSSIBLE AN EFFORT SHOULD BE MADE TO CAPITALIZE UPON EXISTING CONFLICTS BETWEEN COMPETING BLACK NATIONALIST ORGANIZATIONS.*

*[emphasis added] When an opportunity is apparent to disrupt or NEUTRALIZE [emphasis added] black nationalist, hate-type organizations through the cooperation of established local news media contacts or through such contact with sources available to the Seat of Government [Hoover's office], in every instance careful attention must be given to the proposal to insure the targeted group is disrupted, ridiculed, or discredited through the publicity and not merely publicized. [*Following are transcripts of official FBI COINTELPRO documents obtained under the Freedom of Information Act. The March 4, 1968 memorandum was sent out by J. Edgar Hoover himself just one month before the assassinationof Martin Luther King, Jr. It specifically identified Elijah Muhammed and the Nation of Islam as primary targets of COINTELPRO, as well as Rev. King's Southern Christian Leadership Conference. Other released FBI documents show the Bureau had infiltrators within Malcolm X's Muslim Mosque, Inc. and certainly within The Five Percent.].*

Between 1964 and 2014 our worldview changed morally, philosophically, socially, culturally, and politically. We were the youthful generation Hoover wanted to neutralize. All of the slander, libel and yellow journalism written about the Five Percent Nation, our Universal Flag, our supporters and leadership resulted from the above memorandum. All the positive we've done was hardly ever mentioned by the government's cooperating *"established local news media contacts."* As a nation, we were not supposed to be here! But we have made it through 50 years of teaching and living this life and we continue to rise to the top. While some have given their very lives in dedication to this culture since our year one, some have just begun to take their first steps on this journey of self-awareness, community empowerment and universal consciousness.

Allah required each to teach nine students that would multiply into a nation. The number nine in our Supreme Mathematics is *Born*, the complete and perfected stage of something brought into existence. From the beginning, moments arose when the Five Percent faced challenges to their history, teachings, leadership, and legitimacy. But, each time, they were able to meet those challenges head on and become morally victorious in the face of opposition. Allah himself faced forces of opposition waiting to neutralize, mislead, misguide, and disrupt the order of his nation.

Backstory

The time in which Allah and his Five Percenters emerged was one of turmoil and unrest in America. In 1963, President John F. Kennedy was assassinated. In 1964, the Civil Rights Act was passed. There was a widespread anti-establishment sentiment across America. People questioned police authority. In 1965, Malcolm X was assassinated. In 1968, Robert F. Kennedy and Dr. Martin Luther King, Jr. were also assassinated.

On June 13, 1969, Allah was on his way to the apartment building of his wife, Dora, when he was attacked by assailants. He was reportedly shot and killed by unknown men in the elevator of the apartment building. Although there were speculations about who was involved, none of those speculations led to a single arrest.

While many Five Percenters strive to live in accordance with what Allah actually taught, some wonder if the assassination of Allah was about the man himself or his teachings? Are the teachings espoused by members of the nation today the same exact teachings as Allah himself taught? 50 years of living evidence is seen and heard everywhere to show and prove the answer is unequivocally yes, but caution who you're speaking to these days. The statistical application of the 85%, 10%, and 5% tells us for every person who claims Five Percent, there are also frauds that will mislead people.

The Gods who visited Allah while he was in Mattawan disseminated his teachings to their peers outside who desired more knowledge. In this sense, Allah never left us because of that. This helped the nation continue teaching and building. When Allah returned from Mattawan, he saw all these new Five Percenters at the first Parliament in Mt. Morris Park and it was plain to see the Gods were teaching. While there was some dispute whether Allah approved the name Nation of Gods & Earths, the contribution can be attributed to the Gods & Earths within the Five Percent who embraced it because of their growth and development in comprehension about their culture. Just as our annual Show & Prove was not chosen by Allah himself but it is a universally accepted event that commemorates the life and teachings of Allah, the name Nation of Gods & Earths likewise came from the minds of Allah's fruit who began realizing they were more than just the Five Percent, they were realizing they were Allah (God) as Allah seen himself.

Supreme Mathematics, Supreme Alphabets, and 120 lessons passed down orally from the first generation to our generation today. We were taught to take nothing on face value, to examine everything, go to school, get a degree, get a trade, take care of our family and to keep teaching the babies. And whenever we wanted to see Allah, he told us all we have to do is *"come together."* It goes without saying that one of the challenges of coming together, though, is agreeing on what changes must happen in order for us to adapt to the changing world around us.

Some did according to what he taught and some did not. Those that did became the forces that propelled the nation's growth and development. Those that didn't represented the stagnation that needed change. Growth and development requires that change has to happen, that's a natural process. Some people are reluctant to change, holding on to strange, mythological, and antiquated ideas. Others embrace factual and relevant ideas fitting for the times. Varying opinions about a number of issues has left us appearing undecided and somewhat divided which cripples our ability to move forward as any industrious nation. There is a heavy religious-type reliance on what Allah said and what he didn't say, but very few *does what he did* to exemplify what he actually taught. A tree is known by the fruit it bears.

This generation has been fortunate to have the word become flesh as Allah revealed his identity to babies. But everyone that he taught did not share his vision, they were not like-minded, and therefore went separate ways. Allah attracted many but some repelled from him. Some of those who received the word from Allah kept the word pure without mixing, diluting or tampering with it. Some just had to do a remix though, giving people an artificially flavored version of what he taught. Some took a lot of knowledge and wisdom with them to their grave. Some accounts do not agree with each other, leaving many confused and others feeling and thinking, well, a bit religious. Some are not around anymore, some do not remember that well, and let's not forget those whose sole purpose is to ensure we do not grow and develop into what Allah intended. Yes, we know about them, too.

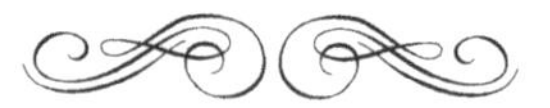

"As life change you change. If you don't change you are going to die. Change with the young people. If you stay around young people you'll stay young."

—ALLAH

Let's recall, he joined the U.S. Army in the early 1950s and was stationed in Korea from 1952 to 1954, where he served as an infantryman in the Korean War. The infantry officer is responsible for leading the infantry, the main land combat force and backbone of the Army. They are responsible for defending the country against any threat by land, as well as capturing, destroying and repelling enemy ground forces. After returning to the U.S., he lived in Harlem and served in the U.S. Army Reserves until he was honorably discharged in 1960. He had foreign service in Japan and Korea, and was awarded the Korean Service Medal with one Bronze Service Star, Combat Infantryman's Badge, Presidential Unit Citation (Republic of Korea), United Nations Service Medal, & the National Defense Service Medal. Shortly after, he joined Temple #7 in Harlem and became an advanced FOI.

After departing from the Temple into the streets of Harlem during the Black Power Movement, the FBI under J. Edgar Hoover, turned their attention to Allah and his Five Percenters. The FBI had a working relationship with the NYPD during this time. Based on a careful review of the FBI file on the Five Percenters, I agree with the assessments made by Bradley R. Goodling. In retrospect of the events that took place after Allah's false arrest in front of the Hotel Theresa on May 31, 1965, Goodling said:

"[S]omeone had made very quick decisions to single out Clarence 13X," and his court appearance *"exhibited signs of railroading."* [*]

After going to the court of his sentencing myself and retrieving official court records, I can say that he was indeed railroaded. He was denied the right to represent himself in court and was falsely committed to Bellevue Hospital on July 9, 1965 on trumped up charges. The way a "system" works is that everyone that works for that system plays a part in a process to achieve a desired result beneficial to the system. You have government officials and courts who work with prisons and hospitals. Goodling pointed out, "He remained at Bellevue for three months without an actual, official sentence from the courts. Then, to add further insult, 'on 10/15/65...Bellevue Hospital officials advised the presiding judge that they needed 10 additional days to complete their psychiatric treatment of subject [Smith]."[†]

The "system" began sharing information per interdepartmental memos regarding the evaluations and hearings regarding Allah. Goodling noted, "While informing Hoover of a hearing for Clarence 13X on the same day, the New York SAC wrote, *'it is expected that when subject appears in above court on 11/16/65, subject will be adjudged criminally insane and will be committed to a mental institution.*[‡] Look at the language they used, "It is expected," "subject will be adjudged," and "will be committed." This language was written by government officials who wrote with clear previous knowledge of their intent and desired outcome.

---

* Goodling, *"Poor Righteous Teaching: The Story of the FBI and the Five Percenters,"* p. 14.

† Goodling, *"Poor Righteous Teaching: The Story of the FBI and the Five Percenters,"* p. 14.

‡ Goodling, p. 15, footnote 102: Federal Bureau of Investigation, "SAC New York to Director," *Five Percenters*, #157-6-34, part 2, November 16, 1965, Ikd. *Freedom of Information Act Reading Room Page*, http://www.fbi.gov, 1

Goodling previously observed the same, "[u]nquestionably, the November 16, 1965 SAC Memo is the most curious document in the Five Percenter files. How could the FBI 'expect' that Clarence 13X 'will' be sent to a mental hospital? One would expect the SAC to use more uncertain, ambiguous words such as 'maybe' or 'could be'; but the words 'will be adjudged' clearly imply some prior knowledge of the verdict. The only way the SAC could have had prior knowledge was if he had some form of interdepartmental contact with Bellevue Hospital, the District Attorney, and possibly the presiding judge. If this was true, then one can ascertain the purpose of interdepartmentalism."*

There were two points Goodling noted that should be considered here. The first was that "[t]he incarceration of Clarence 13X in a mental institution serves as another indication of railroading. Though trumped up, the charges and evidence against Clarence 13X probably would not have landed him in prison for a long time. Smith was a decorated veteran of the Korean War and undoubtedly, this too would have helped his cause during sentencing. However, if Clarence 13X could be diagnosed as having a 'psychological problem,' the law said the sentence of time did not apply. Therefore, he could be remanded to the New York State Department of Mental Hygiene for an indefinite confinement. This was the Bureau's ultimate goal. With Clarence 13X in a mental institution with no recourse for release, he would, so the theory went, be out of the picture." The second point was, "A problem ensnared this scenario, however, because Clarence 13X never had complications with his mind. Based on his original philosophies and accomplishments as a leader, Clarence 13X was a man of high intelligence. Yet, the court tried to pass him

* Goodling, p. 15

off in the November 16, hearing as being 'unable to understand the charges against him."*

In his own words as a survivor of the system's injustice, Allah explained *"They sent me to Mattawan because I said I was Allah. Then they certified me out of Mattawan as Allah. I could have sued this city. But I don't want no money."* As you will see from notes taken from the Otisville interview, Allah was a decorated Korean War veteran who had his faculties intact. He was a leader and not a follower. He went out by himself with inner strength and will, the willingness to write history a year for every mile and he went the extra mile to do the hard job that nobody wanted. He was successful in reaching the unreachable, he taught the unteachable, and he corrected the incorrigible. And of the seeds he planted?

A Seed in Fertile Soil

Everything that lives on Earth has a life cycle. Life begins, it grows, it reproduces, and it dies. But what is a plant's life cycle? Plants start their lives as tiny seeds. Seeds can be as tiny as a grain of sand or bigger than a fingernail. Some are round, while others are flat or pear-shaped. Inside a seed is an embryo, which is a tiny plant, and the endosperm, which are small leaves which supply the embryo food. The outside of the seed has a seed coat, which protects the embryo from injury or drying out.

Some seeds have very hard seed coats. Others have soft seed coats. Morning glory flowers have hard seed coats. Some gardeners soak the seeds in water or nick them to soften the coats so they'll grow faster. Some seeds need cold temperatures to

* Goodling, *"Poor Righteous Teaching: The Story of the FBI and the Five Percenters,"* p. 15

break down the seed coat. This is called stratification. All seeds need moisture, oxygen and the right temperature to germinate (or grow). Until they have these conditions, the seed remains dormant and does nothing. Some seeds need light to germinate. Others need darkness. Once the seeds have the right conditions, the plant inside starts to grow and get bigger. It pushes open the seed coat — sort of like a chick hatching out of an egg. Tiny leaves appear and push out of the soil. Some plants, like ferns, don't produce seeds. They make spores. Look under the leaves and you'll see rows of tiny round spores. These drop off the plant and eventually make new plants.

Animals often eat seeds. The seeds come out in the animal's poop. They drop to the ground and make new plants. Some seeds are carried to new places by the wind. Seeds don't grow well if they land right underneath the parent plant. There's not enough light, water or nutrients there. Every one of us that have children, have what we call seeds. Each one of us, likewise, is a seed that has great potential but needs the proper environment for our growth and development or else we will stay dormant and do nothing. It was a blessing for me to be around good brothers and sisters who set up the best environment for me, I am still growing.

We have and continue to produce a great number of seeds. We protect, nurture and teach them so they can grow and develop to use their power. We don't want their power to be dormant, we want our seeds to sprout and mature. Some of us went through those periods of dormancy but we needed a new location to survive adverse conditions. The first born were mostly teenagers being prepared to be community leaders, some didn't take the teachings seriously, and the conditions needed to be favorable for their growth. According to Brown Seed Shamgod's interview recorded in 2006 with Rasheen Allah speaking on the early Five Percent history, Kihiem [first born] was one of the most instrumental brothers to spread the teachings in youth facilities as a youth himself. *"He spreaded the teachings by spreading the lessons,"* Brown Seed Shamgod said.* This generation is not all that different but the conditions are different for this generation. There are more distractions, the music is different, and people associate more on social media than they do in their own neighborhoods.

Allah was with us for five years, three of which he spent in Mattawan, and only a few people went up there to visit him. Looking back at what Allah said in the November 1968 Otisville interview can give us some insight into his intent for the nation. Only a few writers referred to this invaluable interview as a

* Youtube: *"Brown Seed Shamgod Speaks on Early Five Percent History"*

first- hand account of Allah's views without an interpretation being offered by some quasi-priesthood. There is no conjecture here of what Allah taught, no interpretations, no inter-orientations, no freestyling, no hidden agendas, no one speaking for him other than himself.

Based on this record, it is impossible for a person of average intelligence to conclude he was a so-called *"Black Supremist,"* or a so-called *"hate teacher,"* or a so-called *"gang leader."* These stigmas were affixed to divide and confuse the people. We don't need anyone's permission or approval to build a nation. We don't have to integrate with people who are opposed to our concerns to build a nation. We don't have to argue, fight, or kill each other to build a nation, either. We don't have to compete with each other to build a nation. We don't have to create separate entities or movements to build a nation. But, what we *must do* is come together, plan, organize, and **learn to work together** to build a nation. What we *must do* is learn to agree on common interests instead of differences of opinions. What we *must do* is become proficient in all forms of mathematics, the sciences, the law, government, agriculture, business, and economics. And we *must* have rallies and Parliaments with focused agendas and cut down on the socializing and fraternizing *if we are serious.* That should be left alone until after the Parliament. Another suggestion might be to have time-keepers at meetings so people won't go into personal tirades that are neither well thought out, planned, or have nothing to do with our national agenda. Those discussions can be dealt with after the Parliament.

Allah's work in New York City is a work that is still going on. It by no means ended because one person or certain figures are not here, it didn't end because New York City has a new Mayor, and it didn't end with the drug epidemic. There are Five Percenters all across the globe today from New York City to London. There are those of us who live as far away as China and South America and as close as New Jersey. Regardless to what country, state, city or neighborhood you are from, a Five Percenter is there teaching or working for the greater good of at-risk youth. This is the side of the story that reporters tend to leave out of their articles. As aptly stated by God Kalim, Editor of The Five Percenter newspaper:

> *"Reports of the demise of the NGE have been greatly exaggerated. Certain Gods, Earths, and 5% just won't let that happen. Since most Gods and Earths teach everywhere they go, the knowledge of our nation has spread from coast to coast and overseas. In the early days each of the five boroughs [Mecca, Medina, Pelan, the Desert, Savior's Island] had high tides and low tides. When the knowledge subsided in Mecca [in 70] it flowed in Pelan. If the knowledge subsided in Medina [in 72] it flowed in the Oasis. During high tides the knowledge flowed everywhere and beyond. During low tides the knowledge subsided here and there. No matter how it may look in one location,* ***the knowledge of our nation is always always always growing somewhere.****"*[*]

* God Kalim, *The Five Percenter Newspaper*, June 2014, Vol. 19.10, p. 2

CHAPTER 2

Good God, Almighty! Allah Builds In NYC

A TRUE VISIONARY FOR a new day of peace and building with strong, healthy young people at the forefront of a nation, Allah lead by example. His work with Barry Gottehrer, aid to former New York City Mayor John Lindsay, was an integral part of the Urban Action Task Force's success. It was established to prevent racial violence in the city, and the two worked to create

programs for Harlem's youths. Their efforts led to a relationship with the Urban League and the creation of a street academy called the Allah School in Mecca, intended to help Five Percenters graduate from high school and continue their education. Allah was well aware that racism and prejudice were problems society faced. His moral position trumped the psychological motives and prejudices that were the basis of the racist norms of the day. He knew evil spared no one and anyone, regardless of skin color, could be corrupted.

Heavy drugs flowed into the Harlem community (and many other Black communities) because of the ambitions of Nikki Barnes and Frank Lucas, but Allah taught against the use of dope. They shared the same skin color as Allah but they surely weren't like-minded. During the height of what many have described as the "Black Power Movement," Allah advocated a pro-righteousness approach. In the often-referred to Otisville interview, he explained: *"We might look a-like but we may not be a-like. See un-alike attract and a-like repels. Now un-alike attract is between me and you. If we have the same mind together then it is not un-alike. It is a what? Like!"* Moreover, in defining the moral position of his teachings, he maintained: *"We are not pro-black nor anti-white, we are pro-righteousness and anti-devilishment."* Some are confused about the two positions today as the lines of right and wrong have been blurred. How many think today is not the thinking of God, but thinking from an un-alike world created by opportunists.

The evidence of how drugs destroyed the fabric of the Black and Latino community is overwhelming. I grew up in a lower middle-class neighborhood in Queens that was destroyed by drugs. It was the heroin capital of Queens. Social activities had to always involve some type of shooting dope, sniffing coke, smoking or drinking. These activities, regardless of how harmless they may have seemed or how good they made someone feel, always brought the worst character out of people. Drug abuse affected so many good people in such a way that they became demoralized (liars, cheaters, thieves and murderers) too far gone to completely clean theirselves up. The addict never experiences stable happiness, just a quick fix or momentary escape. While many recovered from this condition, this adversely affected the Black and Latino community as a whole, and specifically conscious movements within these communities including the nation. Many fell victim to drug abuse and the so-called war on drugs sent a great number of addicts and drug pushers to jail. Today, it is evident that drug and alcohol abuse is no good for any person, family, community or our nation.

In the first edition of *The Righteous Way*, I briefly discussed the working relationship between Allah and Barry Gottehrer, a journalist turned political crusader who rolled up his sleeves to get involved in the issues of New York's most troubled neighborhoods. His perspective might be helpful in providing insight on what building alongside Allah was like. He was the exact opposite of your usual all-talk city officials. When Gottehrer joined the Lindsay administration, he reached out to dialogue with New York's Black leaders. Number one on his list was Allah. According to Gottehrer:

> *"At the time I met Allah, his followers numbered either from two to five hundred, if you believed the police reports; or eight hundred, if you believed Allah. I'd been told that sooner or later he could try to start some kind of armed revolution, and that the Five Percenters would try to take over Harlem. But Allah wasn't sounding like a revolutionary. He was asking me if the city could provide enough buses to take the Five Percenters on a picnic. I said I thought it would be possible. He then asked about school facilities. The Urban League, which used corporate funding to staff store-front street academies for dropouts themselves, told Allah they weren't willing to give over a street academy for the exclusive use of Five Percenters. Allah wanted the city to do better by him."**

The NYPD and New York newspapers filled with exaggerated fear and misunderstanding, characterized Allah as a radical, militant leader of a street gang. Meanwhile, Allah was seeking city assistance to provide bus trips and a school for his Five Percenters. According to Gottehrer, *"Allah held that all men were Allah, and had divine*

* Gottehrer, Barry, *"The Mayor's Man,"* pp. 93-94, Doubleday, 1975

*qualities. Through education, these qualities would be revealed. Since each person is Allah, each is qualified to teach as well as learn.** The reports describing Allah and the Five Percenters never added up if you met them in person. They didn't add up to Barry Gottehrer either as he described his first Parliament:

> *"Allah invited me to visit the monthly gathering of the Five Percenters, which they called their Parliament, to explain about the proposed picnic and the possibilities for the school, and I accepted for the last Sunday in May. All the time we talked, my mind kept drifting back to the police report on Allah. It didn't go together. I began to wonder if I was talking to the right man. Allah's sidekick, Jesus, who for some unexplained reason later changed his name to Justice, was a fine-looking man with white hair. He didn't fit either. When I asked how we'd get in touch, Allah gave me the numbers of the two pay phones at the Glamour Inn and I wrote them on a matchbook, promising I'd call him within the week. Then we ended our conversation as Allah had begun in, with a handshake and the assurance, "Peace!" The next few days brought me a piece of luck: I learned that the city owned a building with a store front five doors down from the Glamour Inn. They had taken it over in a tax case, and the store was closed. I called Eugene Callender, who then ran the Urban League (and now runs the New York Urban Coalition), and we arranged for the League to run a street academy for Five Percenters: the city would turn over the store front to the league for a dollar a year if they would supply two teachers. On the last Sunday of May, I was ready to attend the Five Percenters' Parliament, held on the top of Mount Morris Park from twelve noon until evening. The police had told me that the Parliament usually attracted between one hundred and two hundred people. They kept their eye one them, but from a distance. When the group was meeting, only black policemen would be sent anywhere near the area. They would advise against a white man going up there at all. Sandy Garelik was so concerned about my safety, he insisted I walk over there*

---

* Gottehrer, Barry, *"The Mayor's Man,"* p. 94, Doubleday, 1975

with two policemen, both black—one was a plainclothesman and the other was Cal Boxley, a lieutenant in uniform. Remembering my discomfort in the Glamour Inn, I did not wear a jacket and tie, but was dressed in a T shirt and slacks." *

* Gottehrer, Barry, *"The Mayor's Man,"* pp. 95-96, Doubleday, 1975

Barry Gottehrer went on to describe a vivid picture of what our nation looked like in its humble beginnings, some of the ideals and values Allah set forth for us, including how Allah termed our work a "mission," not just another movement, hangout, social club, or a fad:

> *"We went to the park, which interrupts Fifth Avenue between 120th and 124th Streets, and walked up the long, winding circular steps to a clear area at the very top around an old landmark lookout tower, where Five Percenters from all the five boroughs were now assembling. From that high point, you can see all of New York City spread out below. I also saw winos up there, relocated for the afternoon, staggering off with their bottles; and used drug paraphernalia. Allah was up there with his friend Jesus, or Justice; Al Murphy, another man who hung out at the Glamour Inn; and about two hundred young Five Percenters, all of them black. The meeting opened with the ritual greeting, "Peace!" and Allah gave a combination speech and sermon about their mission to take care of their own, live clean, and civilize the world. It wasn't a new speech, for the Five Percenters would chime in with key words and phrases. The young people, who ranged in age from eight or nine years old to twenty-five, with a handful of older people, stood in a circle and took turns stepping forward to make a short speech. Most of the boys wore small African hats that looked like square Jewish Yarmulkas. There were a great many young girls, most with small babies pulling at their skirts or held in their arms. Few were married. Allah believed that one way to inherit the earth was to produce more children than any other group and out populate the competition. I was introduced and stepped into the center of the circle to speak my piece. Lieutenant Boxley and the plainclothesman, whose identity was no secret from the Five Percenters, faded to the periphery. I explained that the mayor was interested in working with their organization and that we would be able to get them buses to go to the beach in Long Island, at one of the state parks. We settled on an 8A.M.*

> *departure time for the following Sunday. I told them some of the plans for the street academy. When I finished, there was silence. No one believed a thing I said. When I went back to the Glamour Inn with Allah, he was polite but I can see he didn't believe me either."*[*]

That following Sunday, Barry Gottehrer kept his word, the buses arrived on time and we did get that school. Gottehrer recalled, *"Allah told me later that the buses for the picnic weren't so important to him; he was more interested in learning whether I would keep my word. On the street, that was all that mattered. You might be a pimp, a thief, out on bail or hiding from the police, but you would still deal straight in the community if you wanted to keep your self-respect, and if you wanted others to respect you. "**Word**," said Allah, pausing for his followers to add the last phrase with him, "**is-bond**."*[†]

* Gottehrer, Barry, *"The Mayor's Man,"* pp. 96-97, Doubleday, 1975
† Ibid, ***emphasis mine***

CHAPTER 3

Allah Saved NYC From Destruction

IN THE 60S when Watts, Newark and dozens of other cities burned and rioted, former New York City Mayor, John V. Lindsay succeeded in keeping the most densely-populated city in America from erupting in violence — because the people knew he cared and because it wasn't the first time the WASP (White Anglo-Saxon Protestant) mayor had walked the streets passionately listening to the protests and concerns of the poor.

Perhaps Lindsay's most courageous act in 1968

when the Rev. Martin Luther King Jr. was assassinated. Lindsay walked the streets of Harlem well into the night, mingling with tens of thousands of justifiable outraged people, and simply repeating the words "I'm sorry" while pleading for calm. While many cities responded to downtrodden neighborhoods by simply bulldozing dozens of city blocks in the misguided urban renewal programs of the times, Lindsay worked to preserve and invigorate the urban fabric of historic, yet crumbling neighborhoods far beyond the influential reaches of Manhattan. Though the press chided him at the time for a comment that New York was a *"Fun City"* at a time of so much economic, labor, racial and political strife, Lindsay was one of the very few big city mayors who were visionary enough to see cities as great repositories of wealth in the form of diversity, cultural offerings, public transit, density and walkability.

Careful observers of the media understand some journalists attempt to be objective by two methods: 1) fairness to those concerned with the news and 2) a professional process of information gathering that seeks fairness, completeness, and accuracy. Although the ethical responsibilities of journalism are high, the ethical heights journalists set for themselves are not always reached. Allah and his Five Percenters were scoops to reporters in the New York Post, The Newsday, The Daily News, and other New York newspapers. Unlike articles in The Amsterdam News, many articles written about Five Percenters always lacked a professional and balanced process of information gathering. Instead, reporters always seemed to get their information from sources that exaggerated stories instead of giving just the facts as was done in the Harlem Six case.

But on April 22, 1968, an interesting article was published in New York magazine entitled *"Special Report: City on the Eve of*

Destruction," by Gloria Steinem and Lloyd Weaver. New York magazine* is a weekly magazine concerned with the life, culture, and politics of New York City. According to the report, the influence of Allah's teachings is one that certainly impacted the life, culture, and politics of New York City. While many journalists focused on Lindsay's administration, Allah and his Five Percenters were in the streets of New York City teaching those without knowledge of self how to be civilized and righteous. This proved successful as people were on the verge of tearing New York City apart after Dr. Martin Luther King, Jr. was assassinated.

Riots broke out in Los Angeles, Washington, D.C., Baltimore, Louisville, Wilmington, Kansas City, and Chicago to name a few. As brave as Lindsay was for showing up in Harlem, he would quickly learn that bravery alone would not be enough to calm upset Harlem residents. Many confused whites have been led to believe that we are some anti-white hate group. This is hardly

* Not to be confused with The New Yorker

the case as Allah's Five Percent was a force in the community to which Mayor Lindsay was able to turn to on the night of April 4th, 1968. In the book *"The Ungovernable City,"* author Vincent J. Cannato, explained:

"Lindsay was being protected by Harlem operators whom he had gotten to know during the previous night walks. One group to which the Lindsay administration turned was Allah's Five Percenters. Barry Gottehrer had spent much time earning the trust of Allah and his men and it seemed to pay off the night of April 4. Barry Gottehrer, a natural worrier, saw Lindsay's light-colored head bobbing above the crowd and thought, "Jesus, this is just the night for someone to take a shot at him." A fight broke out between the Five Percenters and some bodyguards sent over by a Harlem labor leader. Lindsay was jostled and confused. He later told Gottehrer that he mistakenly believed one group was out to kill him and the other out to save him, but he wasn't sure who was who."*

Civil disturbances swept all over the United States and New York City was going to get very uncivilized. According to the report, once Mayor Lindsay learned of King's assassination, he wanted to go to Harlem because he knew a riot would be born in *"the country's oldest, most politically sensitive concentration of Negro leadership, if it were to happen at all."*† As the report went on, Lindsay said *"somebody just has to go up there. Somebody white just has to face that emotion and say that we're sorry."* But the tension was so thick and the country was splitting in two, it would take more than one white person saying they're sorry to quell the pain of a city.

---

* Vincent J. Cannato, *"The Ungovernable City,"* p. 211; Basic Books, New York 2001

† City on the Eve of Destruction, p. 32A

A report entitled "**Summer in our City: 1967 and 1968: Report to Mayor John V. Lindsay**," by Barry Gottehrer, Chairman of the Urban Action Task Force, detailed the conditions New York City government was in during the time Allah worked with him:

> *"Out of the poverty riots and the racial disturbances that have ravaged our American cities during the past four years, two factors became apparent to the city administration in New York as early as the summer of 1966.*
>
> *(1) There was little or no meaningful coordination among city agencies either in planning to head off summer disturbances or in meeting these disturbances and preventing them from turning into riots once they occurred. This lack of coordination at all levels of municipal government and the failure to share information at the top were particularly evident in the relations between the Police Department and other city agencies.*
>
> *It would not be an oversimplification to say that previous to the summer of 1967 the first contact between the Police Department and other city agencies in regard to a summer disturbance would be a telephone call from the Police Commissioner to the Mayor telling him that X number of men and cars had just been dispatched to a certain location and that there might be a serious disturbance in the city that night. What the Mayor did with this information at this point was entirely up to him. Regardless of what he chose to do, he would be reacting to the report of a disturbance without any previous preparation, community contacts or municipal coordination.*
>
> *In effect, in the summer of 1966, the New York Police Department was isolated from the other city agencies including the Mayor and his staff; yet it was increasingly being called on to deal with tensions and citizen complaints that had little if anything to do with police services. Despite a highly successful community relations program in the ghetto areas of New York, the Police Department increasingly found itself on the front line at explosive situations in which it could supply neither answers nor solutions. In New York, as in other cities where the focus of community unrest and tension had shifted from police brutality*

to bureaucratic brutality, the police were forced to deal with ghetto residents angered over dirty streets, dilapidated housing or inferior schools.

(2) Yet equally as significant—and as potentially explosive—was a serious communications gap existing between the municipal government and the residents of the erupting ghettos of the city. In this city of 8 million people, this communications gap was felt two ways—the resident of the ghetto found it extremely difficult, if not impossible, to break through the bureaucracy to receive desperately needed and deserved municipal services and, equally important, the city administration was too isolated and too weighted down in red tape to hear and truly understand the voices of its less fortunate and less articulate residents. This lack of communication exists for all residents in an overly centralized bureaucracy; it is, however, far more difficult for the lower-income, less-educated citizen to get the services due to him than it is for the wealthier, better-educated citizen. And it is precisely the failure of the ghetto resident to get these services that has created and intensified community tensions.

These two factors—the lack of coordination at the top and the lack of communication throughout—convinced Mayor John V. Lindsay that a radically different approach was needed to meet these potentially explosive problems in 1967.

*This approach was announced by Mayor Lindsay on April 12, 1967, with the formation of the Summer Task Force, headed by an Assistant to the Mayor with and immediate access to the Mayor himself and composed of 29 high-ranking city officials-commissioners, deputy commissioners, assistants to the Mayor and representatives of the Borough Presidents. Each of the members of the Task Force was empowered to make immediate decisions for his department."**

Some of the potential problem areas cited in the report included Bedford Stuyvesant, Central Harlem, East Harlem, Bushwick, Brownsville, East New York, South Bronx, Williamsburg, Coney

* "Gottehrer, Barry, *"Summer in our City: 1967 and 1968: Report to Mayor John V. Lindsay,"* pp. 1-3

Island, the Rockaways, Corona, South Jamaica and Fordam-East Tremont.* The Task Force also included people from the community, from the streets, who had a direct link to the Mayor's office. The report noted: *"The Task Force was a mechanism designed to make the local agencies of city government more responsive on an immediate and direct basis to community needs and desires...Just as John Lindsay himself had gone to the streets, just as he had taken his commissioners with him to the streets of our city, he was attempting through the local Task Forces, to bring local agency chiefs, directors and administrators into direct and continuing contact with the people they were paid to serve. The Task Force provided the Mayor with a direct and immediate link to the most troubled and deprived neighborhoods of our city and it likewise provided the neighborhoods with a direct line to City Hall."*†

The Mayor knew that the action was in the streets and he did what he could to be right there to work with community leaders in rebuilding the ghettos regardless of their exaggerated reputations by the press. For example, mischaracterizations of Five Percenters were known by Barry Gottehrer to be completely false. The following is in stark contrast with the media's mischaracterization of Allah and his Five Percenters being an anti-white group. In his report, he explained:

> *"In New York, experience has proven that many groups or individuals who have been labeled anti-white, ready to riot and totally unreachable by members of their own community and the press turn out to be far more constructive and cooperative in face-to-face confrontations and relationships. (The Five Percenters, a group of some 500 to 700 Negro youth who were*

* *Summer in our City*, p. 5

† *Summer in our City*, p. 7

supposed to be violently anti-white but proved to be quite different, is an excellent case in point.)."*

The riot response of Mayors in other cities were usually to call upon the National Guard, but Mayor Lindsay had an Urban Task Force and telephone numbers of neighborhood leaders he could call before heading into Harlem following the King assassination. The Mayor's strategy was to diffuse the growing crowds on 125th Street and diverting them towards 126th Street where he would eventually meet Allah and his Five Percenters. The streets were filled with Black militants, followers of the Black Power Movement, students of Stokley Carmichael, Malcolm X, Frantz Fanon, H. Rap Brown and Malraux. King's assassination was only more fuel to the raging fire in their hearts.

Gloria Steinem's report noted that Allah started *"signaling his followers north to help the de-fusing plan. This was the result, the Mayor knew, of Allah's contact with the Task Force, but the press still equated any militant with "hate whitey."* (City on the Eve of Destruction, p. 32C). It is noteworthy to mention here that the press has not changed much in how they mischaracterize Five Percenters. According to the report, Barry Gottehrer got *"the Times to change a story referring to them as 'anti-white'; a misjudgment made by one reporter, and repeated by the journalistic custom of writing from clips."*† These forms of exaggerations and yellow journalism took place once again in a New York Post article targeting businessman Jay-Z for wearing the Universal Flag of the Five Percenters, mischaracterizing the Flag as a "hate whitey" symbol. It is my contention that the rumors spread by

* *Summer in our City*, p. 19
† City on the Eve of Destruction, p. 32D

tabloid-style journalism muddies public conception and does harm to public safety by spreading fear and false security threats.

The city seemed divided in two (a black Manhattan and a white Manhattan). King's assassination was the fuse and the report described: *"One police misjudgment, one anarchist, or one black-white street fight could set the match."** Allah's neither pro-black nor anti-white position was neutral and was about peace. He instead taught being pro-righteousness and anti-devilishment; in other words, all for what's right and exact, not for what's wrong. Black extremists, as the report described real pro-Black militants, accused Allah because of his peace-keeping position, of *"being in league with whites."*† So, the idea that we are anti-white is purely false and was not the teaching Allah advocated.

How would have New York City turned out without Five Percenters in the streets teaching people? Where would these people be? How many people got knowledge of self, learned the meaning of being civilized, righteous, getting an education, protecting their women, raising their children? How many stayed healthy and strong? How many died as a penalty for not listening? As the crowd chanted *"We want whitey! We want whitey!"* Gloria Steinem wrote Allah's response after a crowd of people were pumped up by a militant speaker before him:

> *"Them people out there are stupid. They're just showing my Five Percenters they are blind, deaf, and what?" "Dumb!" came the chanted answer. "We are the only ones who are civilized. We are trying to save our people's lives. The revolution must come from within. Clean up your homes first. Our job is to civilize the what?" "Uncivilized!" He turned to one member. "If a Five Percenter don't listen, the penalty is what?" "Death!" "There is*

* Ibid

† City on the Eve of Destruction, p. 32E

L TO R: Black Seed Ralik (Dean in front of Ralik), Abeka, Harule, Lashar, Lamik and Kiheen; Spring 1968, Courtesy of C Allah (Fort Greene, Medina)

no teaching in a bar. Bakar Kasim and two sisters got busted over in Brooklyn where they were teaching. Some cops come over messing with them, and one of the sisters bit him on the hand. He shouldn't have had his hand on her, and the man should have took his head! Now, you know we believe in peace, but I didn't say if we are attacked don't fight! You say you are God, and a sister is in jail for biting a policeman on the hand. Malcolm said he'd rather have the women than the niggers. And another thing I taught you was to respect the American Flag. You respect any what?" "Government!" "I'm telling you, my Five Percenters have got to be healthy, strong and good what?" "Breeders!" "But you're healthy all right. Some of you brothers outrun a reindeer or a telephone call every time a riot starts. You ain't ready! You can't fight no guerrilla warfare here, because you don't plan nothing. You have to buy from the white man. Why? Because when his superiors give him the orders, the

> *penalty for disobedience is what?" "Death!" "You're out there looting. My wife had to go downtown to get some milk. Some brothers are looting. Don't say you're a Five Percenter if you gonna do it! You say you the civilized of the world. The white man won't give you this government until you have given your word you will not destroy him."'**

Of all the community leaders Steinem's report mentioned, the most detail was focused on Allah and his Five Percenters' role in assisting the Mayor in keeping Harlem cool. At the end of a class held at Allah School in Mecca (Allah Youth Center in Mecca) by Allah on a Friday night, Steinem reported, *"Five Percenters began filling out past a life-sized photograph of the Mayor and Allah. The inscription read, "To Allah, thanks a lot, John V. Lindsay."* (ibid).

People oftentimes quote Allah when stressing certain points about what he taught, but instances of uncivilized behavior by some claiming to be Five Percenters exhibit the contrary. We know Allah didn't teach *that.* It's necessary, therefore, to point out that Allah set an example for Five Percenters when it comes to being civilized. *"We are the only ones who are civilized,"* Allah stated as previously noted. *"We are trying to save our people's lives. The revolution must come from within. Clean up your homes first. Our job is to civilize the what?" "Uncivilized!"* He

* ibid

turned to one member. *"If a Five Percenter don't listen, the penalty is what?" "Death!"*

How can anyone trying to save their people's lives do so if they are tearing their people down because they don't know something? How can an external change (revolution) come if we haven't changed from within? How can we clean up our communities if we haven't cleaned up our homes first? And how can each one teach one if a Five Percenter (or anyone) won't listen simply because they claim to be the Almighty? Allah also said some people will return home (experience death) to show people everything is real. Have we not seen many return home? Haven't we learned by now that we have to live right and exact to be examples of what life is all about to our children and the uncivilized? How else can we effectively and efficiently save others? Is it done just by simply quoting lessons? We have acquired, quoted and applied much in the way of knowledge, wisdom and understanding in the last fifty years, so what happens next?

CHAPTER 4

To Be Civilized or Uncivilized?

THE PURPOSE OF our teachings is to civilize (or to elevate) us from a state of mental death into the highest level of productive and cultivated thinking. The concept of being civilized is a social construct meaning society decides what is civilized and what is not. This is usually based on cultural practices and norms people within a society agree are morally correct behavior. The second lesson in the 1-14 tells us *"Civilize means to teach knowledge and wisdom to all the human families of the planet Earth."* The seventeenth lesson (Knowledge God lesson) in the 1-40 tells us the meaning of civilization is *"One having knowledge,*

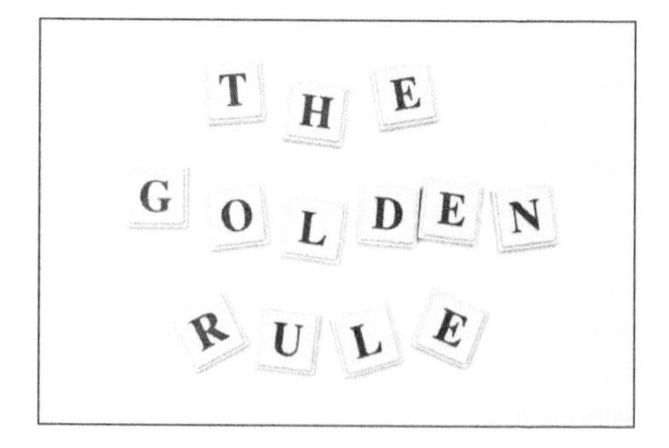

wisdom, understanding, culture and refinement, and is not a savage in the pursuit of happiness." It should be noted that the civilizing here has to take place inside of us first and in our homes before we can civilize other people. Imagine a person who hasn't bathed in weeks coming to you to tell you the importance of taking a bath, or someone teaching your child manners in the street but doesn't teach his/her own children manners at home. As absurd as that sounds to you, some folks do this. So, civilizing starts with personal transformation (mentally) which is compelled by knowledge, wisdom, understanding, culture and refinement, and it flows (behaviorally) outward in how we teach and treat others.

In this context, civilization is a **highly developed** society and **culture**, and one whose present members exhibit **moral and intellectual advancement; humane, ethical, and reasonable treatment of each other.** Our ancestors are the Mothers and Fathers of the greatest civilizations known to man. They originated what it means to be civilized through the arts and sciences. Since many of us pride ourselves on coming from a great history and heritage, we have to make sure our actions today don't disgrace that legacy through uncivilized thinking and behavior. Civilized behavior in this sense is tantamount to righteous behavior. Such behavior, however, is preceded by righteous thinking, righteous attitude (or view), righteous speaking, and righteous actions. This somewhat echoes the Noble Eightfold Path of the Buddha and only goes to show that righteousness is not a new teaching.

The righteous way is a daily, self-saving approach to removing fear and self-doubt so we can realize our fullest potential. In this sense, the Way is ultimately about unlearning rather than learning — another paradox. We learn so we can unlearn and uncover. Here, we uncover TRUTH to find our way. In order to

find our way, we get the knowledge of ourselves and put into practice what we know. It's like learning how to swim for the first time. After we learn to swim, we don't cling to the swimming instructor. We just get in the water on our own and begin swimming. Although we all swam as fetuses, we have to learn how to swim in greater waters. That's what knowledge of self is about. It's about knowing ourselves right and exact so we can practice mindfulness which leads to righteousness. We can think of this as being able to swim 9,000 miles as mentioned in our lessons. It's no use knowing everything about the lessons and not swimming anywhere. Similarly, it's no use knowing everything about 120 Lessons and not allowing the lessons to change your thinking. The lessons are a skillful means or expedient method to changing uncivilized thinking into civilized thinking. They are like a finger pointing at the Sun (knowledge), moon (wisdom) and stars (understanding) — but don't confuse the finger for the Sun, moon and stars.

In the last 50 years, some have fallen into a dogmatic understanding of the lessons without ever changing the uncivilized mentality the lessons are designed to transform. Some have fallen so deep into this *comfort zone* they don't know how to or may not want to get out of it. Others have reached a plateau and aren't elevating any further. The righteous way, therefore, is the path of practicing right thinking and right behavior that can change conditions. I didn't invent the path. It has always been here. It is the way of doings things right. These teachings are leading you in the right direction, but you have to walk the path for yourself. The righteous are like a light to a people who live in darkness.* It's through the example of the righteous that those living in the dark are able to see the light and learn. The word example is defined as *'something that illustrates or shows as a model.'* This means people should be able to see what you teach by YOUR EXAMPLE based on what you SHOW AND PROVE.

We are a civilized people who were once uncivilized. Our civilized position comes with a degree of responsibility to those who are uncivilized because the uncivilized are without knowledge, wisdom, understanding, culture or refinement as mentioned in the lessons. Allah's approach was revolutionary in that he taught knowledge, wisdom, understanding, culture and refinement to young people because he understood young people are the true wealth of a country. By his example, he taught his developing Five Percent they have a duty to teach and be civilized. As a civilized person, he was very much civically engaged. We previously discussed his leadership role in former New York City Mayor Lindsay's Urban Action Task Force. Moreover, his working relationship with Barry Gottehrer led to the acquiring of a storefront once owned by the Urban League which is now Allah School in Mecca. This was a result of his civic engagement and vision for young people. If he was uncivilized, he wouldn't have been so civically engaged. So, what is civic engagement? Civic engagement is individual and collective actions designed to identify and address issues of public concern. Civic engagement can take many forms, from individual voluntarism to organizational involvement to

* *For more about The Darkness, see Chapter 2 of The Righteous Way, Vol. 1*

electoral participation. It can include efforts to directly address an issue, work with others in a community to solve a problem or interact with religious, political, or other civic leaders.

It is a contradiction to teach others to be civilized while we act uncivilized. So, our ways and actions are also a part of our duty. If we are uncivilized, we don't have a duty to do anything. But, as civilized people, we have a duty to act. The term Duty to Act is a legal term that defines an individual or organization's legal requirement to take action to prevent harm to a person or the community as a whole. Therefore, if you say you are down with this and others say they are down with this, there is a moral and ethical responsibility first to themselves and their family and second to the communities in which they live. This brings us to the Golden Rule.

The Golden Rule is arguably the most essential basis for the modern concept of human rights, in which each individual has a right to just treatment, and a reciprocal responsibility to ensure justice for others. All versions and forms of the proverbial Golden Rule have one aspect in common — they all demand that people treat others in a manner in which they themselves would like to be treated: *Treat and do unto others as you would like others to treat and do unto you.* This maxim is the rule or ethic of reciprocity, balance and fairness. The Golden Rule in general serves as a motivation toward proactive action. This concept describes a "two-way" relationship between you and others that involves both sides equally. Those who are uncivilized miss, lose (or forget) this rule in their thinking and consequently, in their behavior towards others. This can involve a person empathizing with others, perceiving their neighbor also as part of "self," and treating all people with consideration, and not just members of his/her own in-group. Are there disrespectful, ill-mannered, inconsiderate, mean, nasty and evil people out there? Of course, but don't give up on your righteousness just because someone gave up on theirs. Being civilized is practing the ways of peace.

CHAPTER 5

Earth's Axis

THE WOMEN, BY virtue of genetics, anthropology and culture, are without question the mothers of civilization. In the first edition of this book, both historical and anthropological evidence was provided in support of this fact. Just as the Earth is a planet abundant in natural resources and is the place of our civilization's birth, so does the Black woman comparatively fit this role in nature and in mind. The Earth's axis remains tilted and because of this tilt, she has her seasons. The seasons

prove the Earth's yearly revolution around the Sun. In this sense, our women are revolutionary. By nature, she is the nurturer of change, hence the name Mother Nature. Mother Nature (sometimes known as Mother Earth or the Earth-Mother), is a common personification of nature that focuses on the life-giving and nurturing aspects of nature as they symbolically relate to the general nature of women. The Earth's (or Original woman's) axis is the straight line of rotation she makes as she applies herself relative to these teachings. Her applicability of the knowledge she obtains is relative to her speed of rotation and whether she'll be able to balance her axial tilt (carry herself well according to the culture). In order for her to rotate (carry herself) well on her axis, she must have access to two things: 1. Access to correct knowledge of this culture; and 2. Access to an environment that allows for self-expression (rotation).

Let's look at these two a little closer. *Access to knowledge* means getting the proper information and/or education from a credible source that teaches without corrupting, mixing, or diluting any part of the teachings with intent to misinform, mislead, mis-educate, or limit a woman for purposes unintended by the lessons. *Access to an environment that allows self-expression* means inclusion in places (for example, at home and abroad) where a woman can express or apply what she has learned to aid in the nurturing of the culture rather than an environment that suppresses or minimizes her intellectual input and intuitive wisdom.

Looking at our history from a matrilineal perspective, there would be no civilization to speak of without the Mothers of Civilization. Our women have nurtured us into the meaning of civilization since ancient times. Like so many women in the nation, the sisters here share their knowledge and wisdom in their own words to benefit the reader. Sisters were not always referred to as Earths, as noted by one of the first women to teach this culture. According to Queen Omala:

PHOTOS: Courtesy of Earthly Paradise

Queen Omala speaks at God Appreciation Parliament in Mecca (2013)
A Divine Photography, Allah Divine

"We were not known as Nurses we were the Moon reflecting the light of the Sun. In the beginning there was much, much unity. We were a close knit family. I have two brothers who had knowledge of self, so automatically my old Earth stopped eating the swine. So Allah came to dine at my home, Kareem's home and I-God's home, he knew that it was clean eating.

*One of the 1st books that we read was How to Eat to Live. Allah gave me my name meaning "Woman Warrior," We were to be prepared at all times for whatever. The name Earth just came about just like the 3/4's. We must always bear witness that time is change, absolutely nothing stays the same."**

Queen Omala's statements stand true. We must always bear witness that time is change, absolutely nothing stays the same. We are growing and developing each and every day, in each and every way whether we notice the change happening or not. On Earth, we have what appears to be night and day and the seasons are evidence we are on the move. The Earth travels on its axis at a terrific speed of 1,037.33 mph. It's so fast we don't feel ourselves rotating but we are on the planet moving at that speed. Likewise, some of us will not see or feel the change taking place but it happens every day. Change does not ask permission to happen, it just does, naturally and subtly, as was experienced by my Earth and first female student to learn 120 Lessons, Lovasia Sunblessed Earth.

---

* Queen Omala — Mother of Civilization, *Early History on the First Earths, '14th Degree & Beyond: A Magazine for the Original Woman'*, July 8, 2008

Earths & Queens at Fort Greene Rally circa 1984. Courtesy of Jamel Shabazz.

MECCA RALLY 2015 (FROM L TO R): Nubia Earth, Original Love Reflection, Queen-I-Bare Witness, ShaliCeea Earth, Beautiful Umi Earth & baby CeeAira Earth, NRich TruePeace Earth, Lovasia SunBlessed Earth, SciDey, Sci-Honor Devotion, Earth Watesha, UmiSudan L.A.W. & I-Queen Divinity

Peace! I was introduced to the Nation of Gods and Earths in March 2012, but did not have the understanding of what "Knowledge of Self" meant until December 2012. Time, as well as a new insightful & dedicated educator led me to finally know the origin of myself and original people in this world. Years prior I stopped attending church because many questions I had about God were not being answered. Therefore, when I was introduced to the teachings of the Nation of Gods and Earths, a lot of things started to make sense and I wanted to learn more.

Learning that the Black Man is God and the Black Woman is the Earth, made me want to learn more about myself, more about history and what the nation was all about; so my journey began. It took me ten months to learn Supreme Mathematics, The Supreme Alphabet and my lessons which consisted of 120 questions and answers that are to be memorized by all of its members as a way to instill discipline of learning and provoke further study. Its goal is to help those with the lessons navigate through life righteously. I wanted to learn 120 because of my eagerness to find previously unknown information that would allow me to dig deeper into the study of myself. I was eager to have a better understanding of me, what was going on around me and the unanswered questions I had about religion (particularly a mystery God). I knew Supreme Mathematics and the Supreme Alphabet were going to be the keys which would unlock a lot of doors. Seeing how the mathematics manifests in different situations and are embedded in this culture showed me I was headed in the right direction; this also motivated me to want to knowledge 120 for self. My educator, who eventually became my God, was another source of fuel in my studies and growth. His main focus was my mental growth and making sure I gained all the tools I needed to help me in my new journey. Seeing your counterpart apply these teachings in the relationship makes all the difference. When we attend Parliaments we can see the strength in the bonds between men and women. Our bond is strong because mathematics is our foundation.

The first Parliament I attended was in Mount Morris Park in May 2012. My first reaction upon arrival was "Wow, look at all the beautiful black conscious people." I thought I was now "home" because I was amongst brothers and sisters who wanted a change for our people and wanted better, as I did. Meeting and interacting with other Earths was also a great experience. I am a very shy individual, yet friendly and I was unsure of how or even if I would be welcomed. It took some Earths a little longer to warm up to me because they needed to make sure I was sincere. Other Earths felt my warm energy and embraced me from the start. Earth's that live and love this culture are very vocal about these teachings; I listened attentively and absorbed their wisdom. Learning how so many characteristics of the planet earth lies in us, learning the importance of natural substances and ways of life are not only beneficial but necessary; for ourselves, babies and families.

It is beautiful when a culture is filled with women and children. It expresses life, growth and longevity. The number of women must increase, for we are the reason the nation will survive, continue to grow and be fruitful. Witnessing the lack of women and babies attending the rallies/ parliaments is heartbreaking; a nation cannot multiply without them. When I look at the history of the nation via old photographs, the rallies and parliaments were filled with women and babies. The first question in my mind was "What happened?" We need to get back to that. Women and babies have the magnetic to draw other women and children. No woman or child wants to attend an event filled with the majority being men. They want to be able to connect with their fellow sisters with KOS. Sisters without KOS will not be drawn to this culture if they see a lack of women and children. They will think this is a chauvinistic culture in which the only focus is men. That outlook is very repelling. Today, what women want is a unified family, a unified culture, a unified people and we have to set the example. Children want to play with other children; teenagers want to connect and speak to other teenagers. Who can relate to them better than their own peers? There are many distractions to steer them from going the righteous way. We need to understand that times have changed and we are not in the same fight that we once were in the 60's and 70's. The children of today do not know what it is to struggle; to fight for everything in order to survive. So we must adapt to the times to make this culture more appealing and intriguing and teach them that this fight is not over. We are still fighting for Freedom, Justice and Equality. Networking with the youth on the street corner of yesteryears might have worked back then, but today we live in a different time; we are living in a world of technology. We have to utilize different sources to connect with our children as well as our sisters. We cannot forget that our sisters are the ones who are raising our babies, so it is imperative that they learn their origin, history and these teachings, so that it can be passed down from generation to generation and we do not go into extinction.

How brothers manifest their righteousness will be the magnetic to bring more women into the culture! God's must stand out from the brothers without KOS. They cannot play the same games that the average man plays on women, because nothing will make you different except Allah/ God being in your name. The main difference between a man without KOS and a man with KOS is that the man with knowledge becomes civilized and is not a savage in the pursuit of happiness. God of the Universe is not just a title; it is a name you must live out and uphold to the highest ability. That will make you stand out; that will make your magnetic stronger; that will gravitate sisters to our culture. Let your knowledge bring forth your actions so that she can get an understanding of what this culture is all about. Show these women your supremeness by drawing them mentally and not physically. This is why women are quick to vocalize independency and society has brainwashed them into believing that they do not need a man in the household to survive. They no longer see the importance and the pivotal role the man plays in the universe, the home of ISLAM. As God, you must teach them the latter. Be subtle in your approach, be patient, take the time to do the knowledge so you can understand their experiences so you know how to handle the situation effectively. This Earth or potential Earth might not have the compatibility you're seeking, and you know this because you took the time to know her mentally. The most important thing is that you have done your duty by teaching her. Some women are looking for a foundation, not to just be courted. They are looking for longevity in a relationship and a family based man.

Any new sisters coming into the nation, I encourage them to keep building. Keep being the positive example of what it is to be a true and living Earth. Continue to be the change that you want to see. Continue to build on mathematics and make sure that it remains your foundation. As the Mothers of Civilization, we must come together as one.

Peace to my Beautiful Nation!
Lovasia Sunblessed Earth

As evidenced by the Earthly voices here, teaching and building are not gender-specific. When everyone can bring their views to the table, we get a whole picture of reality to look at. Knowledge of Self empowers men and women in similar ways: We both inherit a framework that provides a foundation for self-empowerment, self-discovery, and social equality. Self-knowledge is essential in figuring out what's optional and not optional to our individual well-being and this sister explains some of the reasons why. Like many women here, her love for these teachings is reflected in her enthusiasm to build. The experience of self-discovery is undoubtedly a peak of its own. Most knowledge comes gradually without immediate impact. Sometimes, it takes time to grasp the understanding. Like so many sisters in the nation, moments of true realization causes a release of transformative force and energy to be more and do more as this sister, Original Love Reflection, expounds on her transformation:

Knowledge is INFINITE, Wisdom is the WAY, Understanding is LOVE.....

Peace! My name is Original Love Reflection! I am from Why Cipher (Yonkers), Now Why. I am here to share with the universe, what KNOWLEDGE OF SELF, has done for me, and how it has affected my existence as an Original woman. Knowing that the BLACKMAN is God, and that he is the FATHER OF CIVLIZATION, erases from our 3rd eye/ mind, that a deity, spirit or spook in the sky will bring and or provide you with food, clothing and shelter. When you have knowledge, you are awakened to certain facts and information that cause you to become conscious of yourself and your surroundings. Knowledge can be obtained from your experiences in life, your education and or learning. When you come to the realization and

fact that Original people/Black people are the dominant force on the planet Earth, you have now unlocked a new found power and confidence that, in other words, you would have not been able to grasp. As original woman/mothers of civilization, it is crucial for us, to understand our strength and place in the cipher. We walk beside descendants of Kings! We walk beside GOD, the original Blackman! When you walk that walk, and carry the title of Earth or Queen, it is imperative that you have the understanding of WHO YOU ARE AND WHY!! If you do the knowledge to the original woman throughout history, you will be able to discover strength within yourself.

Evidence of the reign of the Original man and woman is forever plentiful. To this very day, in Egypt, you can see the remains of Queen Hatsephut's temple. Most of the information about Queen Hatshepsut, came to light through hieroglyphics and artwork found in the walls of the Queen's temple ar Beir-el-Bhari. The temple itself is a testament to the accomplishments of Queen Hatshepsut. Other accomplishments of Queen Hatshepsut include her organization of ***a journey to the Land of Punt****. The purpose of the trip was trade and evidence indicates that Queen Hatshepsut managed to bring back numerous precious and rare articles back to enhance the wealth of the Egyptian nation. According to historians, she ruled for about 15 years, until her death in 1458 BC, and left behind more monuments and works of art than any Egyptian queen to come.*

Let's fast forward to the 1960's Civil Rights Moment, a powerful sista like Angela Y Davis a political activist, scholar, and author. Prisoner rights have been among her interests; she is the founder of Critical Resistance, an organization working to abolish the prison-industrial complex. Presently, in the white house, you witness, the beauty, intelligence and grace of our first lady Michelle Obama. She is not only the wife of President Barack Obama; she is a lawyer and writer, and the 1st ORIGINAL WOMAN to become First Lady of the United States. Raised on the South Side of Chicago, Obama attended Princeton University and Harvard Law School before returning to Chicago to work at the law firm Sidley Austin, where she met her future husband..

Last, b.u.t. never least, I look to my ol Earth, my mother for love, inspiration and strength. The love, grace, morals and jewels that she has and continues to give me over the years has been more than I could have ever hoped for. She is the one who taught me how to love others, and most importantly, how to love myself. LOVE, is the highest form of Understanding, and Understanding comes in time!!!!!! This type of growth and confidence is not an overnight process.

This goes into my coming into the Nation of The Gods and Earths. I obtained Supreme Mathematics and Alphabet, in the year, 2005, and knowledged the 1-10 & 1-36, however, interestingly enough, it was not until 2012, that I was fully able to comprehend, commit and understand what it was to be an Earth. For me, it was always in my nature to rebel... lol, which makes me so stubborn and outspoken. In truly embodying what it means to be an Earth, it took a lot of self-evaluation and I had to make a lot of changes. One of the changes I made was my dress/ attire. I finally understood the concept of 3/4th's. When you carry the title as Earth or Queen, the way you look and present yourself to the world is vital when you are demanding respect from others. I started to grow out all processes and chemicals from my hair and went full on NATURAL. Once I understood, the mental damage of what altering and chemical damaging my hair, in order to fit into a European interpretation of beauty, I no longer had that desire to "fit in" with the woman on Tel Lie Vision (television). When I was able to build with the sisters in this Nation and see their 3/4ths attire, I felt a belonging and a love that only another sister in this Nation would be able to fully comprehend. I am on the road to knowledging 120 lessons, currently on the knowledge degree in the 1-40. I realized how essential this was for me to continue on into this Nation and also grow as a human being. When doing the knowledge to this history, I take the information and adapt it to the current times that WE as original people are living in. In this Nation, this mighty Nation of the Gods and Earths, we as woman, must realize that we too have to be educated and are directly responsible for raising the babies. It is thru us, that the original man is physically born into this world. It is essential for is to have a proper understanding of nutrition and personal health. Living mathematics is like breathing. It becomes a part of your mental and physical being. Clean living and a healthy life are fundamental to the growth and development of a successful Nation.

There are two pivotal moments for any woman coming into this Nation: the 1st one, of course, is when you meet your God/your Sun in your Universe. The 2nd, is your first encounter with another sister from this Nation. Some of my first encounters were not always positive. Depending on the place and time, human beings tend to be cold towards one another, in general. My philosophy has always been to treat people how you want to be treated. This of course, was and is not always the case. I am so grateful and humbled by the bonds that I have formed with the Earths in this Nation. Yes, it was GOD who gave me the knowledge, however, it was the Earths that encouraged me on my road to KOS and showed me the essence of this culture. In this Nation, we have many successful, intelligent, wise and beautiful women. We are all sisters in this knowledge. It takes someone with confidence, emotional toughness and love in order to be a part of such an opinionated group of individuals. When dealing with conflicts, here are some steps that can be applied in order to resolve an issue in a civilized and mature manner.

- ***Talk to people instead of about them***
- ***Be a problem solver not a problem evader***
- ***Develop a communication style that focuses on future problem solving***
- ***Problem solvers deal with issues, not personalities***
- ***Avoid forming "enemy" relationships.***
- ***Invest time building positive bridges to your difficult people***

A problem that plagues our community is insecurity and lack of confidence. Some of these feelings can be a result, of your encounters in the everyday world or past experience and abuse (physical or mental). That's why it is so necessary, that WE as original people support and encourage one another. Admire and aspire! There is no room for jealousy and comparisons. Comparisons = self-sabotage. Comparisons kill confidence because they put our self-worth on a measuring stick. When you have a full understanding of your origin, your self-worth and your potential, you will be able to give love to others, because you in fact, love yourself. Knowledge of self is empowering!!!

The integrity of the sisterhood is another aspect of the nation that many sisters are interested in and are building to improve. The *Peaceful Queens* group, founded by Victorious Lanasia Earth and co-hosted by SciHonor Devotion, holds a biennial reunion retreat to empower Original women to build companionship amongst other conscious sisters. By building bridges, sharing ideas, recipes, experiences, and words of wisdom, these Queens are able to connect intellectually, emotionally, and culturally. Earths are sometimes closer to each other than some of their own family members. When such groups or bonds between sisters are absent, the sisterhood is at risk of falling apart. This can also have a collaterally negative impact on family life and nation life.

> ***"Women who understand how powerful they are do not give into envy over meaningless things; instead they fight to maintain the beautiful bond of the sisterhood. These are the real women who know that we need each other's love and support to survive in this world. Love is the essence of being a woman. We must be that light of love that seals the bond and unique beauty of our sisterhood."***
>
> ***-BINDU***

True sisterhood cannot be forced. It is has to be developed with interest, patience, reciprocity and sincerity over time. You may have to go through some things together first. Not every woman will be your best friend, nor should she be invited to be in your inner circle, but every woman is deserving of your respect and support when you are able to provide it. Being your sister's keeper should be a natural reflex. It should be based on how you would want to be treated if you were walking in her shoes.

Sisters/Earths network at the *Herban Mama Soul Lounge* hosted by Beautiful Umi Earth (center)

Sisters at the Million Women March in Philadelphia (1997) Jamel Shabazz.

Queen Divinity, Samiyah Radiant and QuintIssential of Divine Cee (Washington, D.C.)

2015 Peaceful Queens Retreat. Divine Equality Righteous, FadPhotography.com

Sisterhood knows no boundary, no class or geography. Sisterhood transcends and it transforms our sisters for the better. Sisterhood is from the heart.

As for fashion, many people like the style of the Earths — the colors, the headwraps, the dresses, etc. Back in the days, Earths "made" their own clothing to fit their style and shape. They designed it, got their own fabrics, and went to work on sewing machines. It always beats going out to buy clothing that doesn't fit you. Imagine how confident and disciplined women who made and wore their own clothes had to be. The way a Queen or Earth looks and represents her self is always carefully considered and was equally rewarding. Sisters saved plenty of money and looked better than what was on those mannequins. If a sister didn't know how to do this, other Earths were willing to help. Without activities like these, sisters loose the opportunity to come together and bond. Like many sisters, Queen TruEarth, a positive sister from California, always wants more Earths here:

I learned about the Nation of Gods and Earths in 1992. It's kind of ironic because that's when the Los Angeles Riot took place after Rodney King was brutally beaten by LAPD Officers. Change was inevitable for me. In Los Angeles (Love Allah), CA, the Gods had Leimert Park on lock. It was a beautiful sight to see back then. The Gods would be ciphered up building on the day's math and how they interpreted 120. I remember them running the Hebrew Israelites from the Park with mathematics-they were so swift. There weren't any Earths at the time and the Gods in LA were young (in their early 20's), so I would sit on the sidelines and watch the brothers build. Gods didn't really build with another Gods Queen and vice versa so I just sat on the bench doing the knowledge. Sometimes Gods would bring Sistas around,

and I'd be hopeful that they would also become Earths. It was always short-lived. While I was living righteous, and eating the right foods, I wasn't ready to make the commitment to be a part of the Nation of Gods and Earths, but I knew I was drawn to it, and I respected what they taught. In 1995, I decided to begin studying 120 lessons. It was difficult because I didn't have a female perspective in understanding the degrees. It wasn't like the East Coast, where there were many Earths where you could see the culture through how they spoke, dressed, and lived. I didn't have that. My seeds' Father was my enlightener and I appreciated the knowledge I was receiving, but I longed for sisterly wisdom. I know when I first got a hold of the lessons, I read through all of them trying to find the parts that talked about the Black Woman. I was disappointed greatly. There was talk of Wisdom, and Queen in the Supreme Math and Alphabet, and then there was 6-Equality that when built upon by the Gods spoke of a woman's limitation. For me, it initially felt chauvinistic. Then as I ventured further, the lessons spoke of useful land, and square miles, actual/solar facts and geographical facts about the Earth, MGT, and GCC, etcetera, etcetera. I was definitely not feeling the lessons. I was feeling as if the Black Woman was not important in 120. I still studied, learned how to wrap a head wrap, and everywhere I went, the only reference point people had was Erykah Badu, so that's what folks called me. I wasn't mad, they were on the right track in their understanding and I definitely was growing and reading more; learning more, and once I began building with other Sistas after finding the yahoo group Peaceful Queens that my Sista Lanasia birthed, I realized that references to the Black Woman were all through-out those degrees. I also realized that I could apply and relate to all things in regard to self.

Determined, I named myself Queen TruEarth. A Queen is a woman of power, strength, beauty, refinement, and style. She is first in rank and power and is a true reflection of God's greatness. She is a ruler of herself, and her destiny. She is the wife of a King. She is spoken of highly and has class and value. She is intelligent, teaches others by example, and exudes confidence. TruEarth meaning true, honest, 100%, accurate, non-disputed, and factual; Earth-a Black Woman who knowledges God and is the receiver and bearer of life and light, and understands that her duty is to teach, nurture, protect, and elevate our children. Yes! I was a TruEarth and I represented all of that.

I continued to move through the lessons, forming my own understanding as an Original Black Woman and what I realized was what all who venture upon the journey of knowledging self realize, this journey initially is only about you-whether you are a man, woman, or child. Knowledge of self is literal. How you see the math in relation to your existence in the universe is your understanding alone. There is no right way to interpret how the supreme mathematics applies to your life. As I went through the degrees I felt empowered. I was growing, getting taller mentally and physically and my thirst for knowledge outside of 120 grew. For me, mathematics was and will always be my foundation-it is my beginning. I give ultimate thanks to the beginning-meaning Allah in the physical and mental. What I got from it catapulted me into a building frenzy. I wanted to DO more; to become more. I wanted to show and prove the first degree in the 1-10. I wanted to make sure that I wasn't a lip professor (I resented those types), but that I lived out that very first lesson every single day of my life. Who is the Original Man/Woman? The Original Woman is the Asiatic Black Woman, the Maker, the Owner, Cream of the Planet Earth, Mother of Civilization and Divine Being of the Universe. I understood my own worth and power in that degree. Since I am a reflection of the Original Black Man- I realized that I as a Black Woman also had to be divine as well. I realized that every single degree in 120 was applicable to me and with that, I was off! You couldn't hold me back, I couldn't ever be limited again, nor placed in a box, you couldn't tell me nothing about not having a mind; none of that was applicable to me anymore. I was all about showing and proving, doing my inner work-educating myself, maintaining and teaching my own 8 babies, and being that Queen to my God, while building brothas and sistas in my community and abroad. I was unstoppable. Still am!

The Earth-The Black Woman is a complex, dynamic system that can never be fully understood. We are comprised of complex and diverse components that make us special and unique in character and form. We are swift, changeable, and responsive to natural and human-induced changes and an Earth ALWAYS improves the climate/cipher she enters. The Earth is resilient and has a powerful magnetic field that at our core protects us from the undesirable solar winds that can disrupt our mental. As Earths we rise above! We are life givers and sustainers. Our importance in the universe is immeasurable and beyond compare. We are phenomenal, amazing, one of a kind, and most necessary in all that is living. We offer safety, support, encouragement, love, and nurturing to both man and child consistently. Through all of life's natural catastrophes, a true and living earth remains grounded in who she is and brings about balance as she delicately spins on her axis at 1037 1/3 miles per hour moving around the Sun at 67,000 miles per hour.

If I can tell Sistas a few things about being an Earth, I would say, "You must build with other Earths." I would ask Gods to allow those Queens that space and freedom to do so, as it will only enhance her as a Queen. It's not easy making that transition from regular Sista to Queen, to Earth-we often have to graft back so to speak and in that evolutionary process we NEED other Queens around us. We NEED to bounce ideas off of righteous Sistas, we NEED to be understood by our A-Alikes and we NEED to understand them, as we are also reflections of our Sistas because the Moon reflects the Earth as well. I would also tell younger Sistas get this knowledge for you; not because a God is trying to sleep with you, not because you are drawn up by his beautiful, warm magnetic appeal, but because you want to find the divinity in self so that you become an asset to humanity. Sometimes, you will be more serious about getting knowledge of self than even the God who is drawing you up. Go with it...the point is to save lives and you start with self. Sometimes, you will be so on point with how you live out your mathematics, it may repel him. Go with it, the point is to save lives, and you start with self.

Also, this way of life wasn't forced on me, and I don't agree with forcing it on anyone else. I don't make my own seeds study the lessons. I offer them the degrees and whoever wants to study will sit down with me and the others and study-some do, some don't. The chosen will step up and get this knowledge for themselves in their own good time; or they won't. Also, once you get knowledge of self, you aren't better than anyone else. When one takes on that mentality, they lose sight of who we are supposed to teach. I have no desire to stand in circles quoting degrees, never have. While impressive that one can memorize all those lessons and I have, I find the ability to put them into practice much more inspiring, exciting, worthwhile, and beneficial to us as Original people. Once you learn how to tie that head wrap and put on that long skirt, that doesn't officially make you righteous-it's part of a larger process of transformation and teaching. Your ways and actions should always be what people judge you on in the end. How you live each day will send a resounding message of who the Gods and Earths are. Everyone has a role. I found mine through Wisdom-applying what I know to bring about an understanding. You have to find your way-that's the beauty in knowledge of self and if you are consistent and true, you will.

I will that Gods and Earths understand our power as a unit-man, woman, and child, our beauty, our diversity in how we express our culture, and our ability to grow, change, and progress for the betterment of our future-the babies.

Peace
Queen TruEarth
He/Her Divine (High Desert), Cee Allah (CA)

Sisters who are new to the culture "must build with other Earths," as aptly stated by Queen TruEarth. Women from different backgrounds should communicate, bond and learn what this culture is really about. It's a blessing for us to have a common language of peace and righteousness in this culture because it unites us wherever we were once divided. The Latin word *"soror"* (as in sorority) means sister. Sometimes, sisters can think of themselves as if they are in a sorority in this school of divine thought. Our culture has a rich history in uniting Black and Brown peoples. Earth Izayaa Allat, a Latina sister looked at the multi-ethnic diaspora within the meaning of Blackness as it relates to sisterhood. Since our culture unites instead of separates Original people from multi-ethnic backgrounds, she shows a person's heritage can be embraced without denying they too are the Original Black man and woman.

SISTERS IN BLACKNESS

By Earth Izayaa Allat

When I received the knowledge of self from my enlightener or educator, Sunez Allah, one of the tasks I was asked was to show and prove I was Original. Being able to track my immediate genetic lineage to the Northern Region of the Dominican Republic was a start. I knew I was Original because the culture I grew up in was a conglomeration of West African, Indigenous, and Moorish traditions. However, that just explained the environment I grew up in but didn't provide proof on how I myself was Black. So, I got deeper and studied some biochemistry of the Original man and woman. I then thought about how this science related to the concept of sisterhood.

Sisterhood has been a topic extensively discussed both in physical Earth Ciphers (gathering of Earths in their particular regions), retreats, civilization classes, and in groups pertaining to the NGE in social media. In all these spaces sisterhood has been discoursed as a social science, particularly examining the behavior of an individual, lifestyle, relationships and internalization of the lessons inside and outside the nation. We are a very young nation, only 50 years since Allah brought this knowledge to the streets of Harlem, and even in this short time it has reached many so-called Latinos and Asians who are living out the culture. It is therefore, right on time that sisters are discussing these aspects of Sisterhood. However with the knowledge of self comes a deeper analysis that must be combined with the Supreme Mathematics, Alphabet, and 120 lessons to truly build on this aspect of womanhood.

Within the spectrum of Sisterhood must be included the science of Blackness. How can we start to build stronger bonds with each other as sisters by analyzing Blackness and taking it from knowledge to born both in a mental and physical way? Learning it and living it?

What makes a woman an Earth is not whether she can sew, cook vegan food or have a natural childbirth, for there are already civilized women living this way. It is that within all these aspects of MGT (Muslim Girl Training), we have the principles of the Supreme Mathematics to guide our reality as Creation living in balance with our Creator, the Original man. Meaning, we use mathematics to verify that what we do stay in accord with the universal laws that show our nature. This requires scientific reasoning and not emotional pleas. In this way, sisterhood in its deepest sense is built because our studies (1) and its application (2) provide us with understandings (3) that are similarly lived out (4) by the women involved. Other forms of sisterhood bonds are formed because there is a common interest in living by learning. The power (5) of our way of life is seen by sharing (6) how we each live our truest supreme reality (7) as the true and living Earth. From this we take the best part for what suits best at that particular moment in our lives to continue building (8) and creating fully (9). Ultimately, these all become different expressions of the mind in the physical form. Sisterhood should serve as a means for sisters to support each other in building. In our reality it could be teaching each other what we are the best knowers at so we can bring these skills back to our homes to enrich our lives.

All people should always have a purpose for coming together. In this case, having this knowledge, it should be to build, not just talk and hang out. I remember meeting, in the physical, Divine IzEarth from Louisiana and she told me a reason why so many sisters' bonds don't hold is because mathematics wasn't the foundation of their relationship. This is where gossip and 85% ways we pick up in our oppressive realities come back to manifest. There has to be a purpose for coming together, Sisterhood must have an agenda other than sit around and let's chit chat about our emotions. We are mathematicians and scientists and these processes of manifesting thoughts have to be applied to our culture.

For example, a young sister, who is Earth, studying to be a Doula, will have to deal and address the subject of abortion. Many civilized women without this knowledge do advocate abortion because they firmly believe it is a woman's right to choose. However, we have freedom as an insight to culture in mathematics, so we know that the freedom the white man's society advocates is entirely different to the freedom promoted through righteousness. As an Earth doula the 85% woman's conditions must be understood. Was she raped or a victim of incest? Then in these cases, the righteous act may be to support abortion and support her in healing from the experiences. Is she being socialized under the Eurocentric framework of thought where their careers are more important than growing into full holistic womanhood? Then she may need (if she permits) to rethink society on a much deeper basis and how she is a pawn in it. In addition, our lessons say that the 85% are led in the wrong direction, are not civilized, habitually savages in the pursuit of happiness, usually poison-animal eaters, who are indeed mentally dead. Armed with this information, Original women are big victims of the larger framework of the Eugenics agenda, prevalent within mainstream culture and masqueraded as "woman's right to choose." It often gives them a sense of power (really the only power they have is to get the abortion) to counter 85 men exerting their own misguided patriarchal power and they have race/ethnicity, class working against them in all of society. With this in mind, the Earth Doula understands a woman that doesn't have Knowledge of Herself. By using Supreme Mathematics through her support of that woman, she can guide her choices as righteous as she possibly can. In this case, the Earth Doula uses the principle of 4-Culture/Freedom to navigate through the myriad of issues of women in the 85 world.

As in the case of the Earth Doula, the application of mathematics and scientific reasoning guides and sets the path of our life. In this same way, relationships between the women in the nation must be built. The elders taught that Blackness is expressed in 16 shades of Black. This meant that our people express Blackness in endless phenotypes. Scholars like Jewel Pookrum and T. Owens Moore have provided information on topics like melanin, which is a key element in our biochemical makeup, allowing us to understand Blackness within the Original man and woman. This research (1) when contextualized (2) can provide insight (3) to all the Original peoples around the world beyond the so-called Negro in America. For the Earths, since phenotype is the expression of our genetic makeup and we all get judged based on how we look, it is imperative we learn and understand topics like melanin so that we can build genuine sisterly bonds that go outside of our immediate "racial" context (i.e. a so-called African-American Earth and a so-called Puerto Rican Earth). We must be educated on topics of race and gender and its subtopics, such as hair, facial features and body types. The interrelations of each at the local level, historically and currently will clue us into understanding each other more fully.

Using the information as well as our intuitive knowledge will allow us to navigate and naturally find those sisters coming in the name that are indeed Black. Barriers would be toppled down and true sisterhood would be built without the constraints of misunderstanding. We would understand that straight hair and narrow features is just as Black as the coarse hair and wide features of other Original women. Social Equality between sisters won't be as awkward because we would understand the quirks of a sister coming from a different social context. Nina Simone's song "Four Women" is a great example of the particularities in experiences of four different shaded women who are all Original. Just relating to the African American case, the Spike Lee joint "School Daze," depicts the issue between sisters of lighter and darker shades through the sorority rivalry. We also see these issues come to light through the dating choices some of the brothers make, such as the pro-Black militant character truly being questioned for his dating a dark skin sister. Today, we continue to be plagued by these social circumstances.

Sisterhood in its most revolutionary sense, I envision Earths that are bonded by the thirst to learn, that understand each other's immediate history (you can take a step back and truly learn what social context each sister is coming from), the various sciences and come together to focus on enriching themselves through building where divisions of emotions will seize to exist. It is imperative that we move beyond the exterior development of "fashionista" attitudes of what constitutes the Earth. Once we know how Black and Brown were separated to graft, then we can unite them piecing together our ancient history, our current reality applying them to our understanding of Supreme Mathematics, Alphabet and 120 lessons. Sisterhood must be an awareness of Blackness of all shades throughout the world, united and propelled by their knowledge of self. No other conscious circle has ever fulfilled such a standard. Understanding Blackness is the start to deconstructing our perceptions of each other while simultaneously coming together as sisters, the daughters of Mother Earth, Creation whose Creator is Allah, the Original man.

Peace,
Earth Izayaa Allat.

There are many communities in which Black and Brown live but do not have sister circles that creates bonds between them. For women who are Five Percenters, that problem is solved by not only learning the Knowledge of Self, but also having a common cause that brings women together. Knowledge of Self is a powerful learning process that once that journey begins, you no longer will let anybody falsely teach or tell you who you are. More than half of Latin-America is comprised of Original (or Black) people, but asking a person what they are depends on how much they really know about themselves and their history. Being an Original woman who is Latina didn't stop *E-Queen Glorious Earth* from learning the knowledge of herself either:

Coming up in the Bronx where Latino music vibrated out of car doors and apartment windows was far from the culture I was exposed to as I became engulfed in learning more about myself. Beginning with developing understanding and uncovering the truth of what beautiful was...gaining the understanding that there was more to a woman/female beyond her physical attributes...that respect comes greater when you value yourself more. From there, the skies opened up for me. ... Who am I? From where exactly do I come? Who were the "Great Ancestors"? And, where am I to go? What were the Nation of Gods and Earths (NGE) going to do to satisfy my quest of getting to know myself?

Being amongst brothers and sisters who were already endowed in this life-giving information, and having a partner who blessed me with such jewels, he helped me to soon recognize I had come from a history so vast and influential and that the people from my family's homeland, the Dominican Republic, were Black people of so called African descent. From the multi-complexioned skin tones of light, bright, damn near white to blue-black melanin, to the rhythmic movements of Salsa, Merengue and Bachata, to the 1522 Muslim Slave Revolt in the Dominican Republic, to the Orishas such as Sango, Oshun, and Obatala of the Yoruba Culture practiced in Cuba and Puerto Rico, to the Pyramids and Africoid features of the Olmec Stone heads in Mexico, "Latin" America's Black history and "African" lineage is undeniable! All along, I was disconnected from my African Heritage, taught we were different, made to feel inferior when in reality "Latinos" or "Hispanics" were also Original Black People whose history had been distorted. The only thing more powerful than the NGE connecting me to my Blackness is the same teachings that from the Quran, to the Bible, to the so-called Egyptian Book of The Dead, all the way back to the Ancient Kemetic scripts and Metu Neter of our ancestors that have always taught, is that the Black Man is God; the personification of the Primordial, Infinite Dark Energy that always existed prior to the known universe and that the Black Woman is the womb of God, the personification of Dark Matter and Blackness of space that gave birth to the sun, moon and stars. Having been exposed to the teachings of the 5% Nation of Gods & Earths allowed me to delve into a culture that represented strength and power, accountability, intelligence, respect, royalty, integrity, and family. It was a source for me to have the aptness and character to walk with confidence after recognizing that which we were, Gods & Earths, Kings and Queens.

Being a part of the Nation facilitated so many positive choices and decisions because it was a platform to build upon. Yes basics got covered, getting to know self...history...but understanding the present was like rising to the sky without limitations. Natural living became more of a reality to stop going against the nature of Universal law. My diet changed and throughout the years transitioned through a vegetarian lifestyle, embracing Veganism, and currently living off of a completely plant based diet so to eliminate any Western medical interventions. Prior to conceiving, the decision to homeschool became our family's approach towards education. The benefits of homeschooling for us were to target open minds, to raise them according to what will suit their existence in its present and future environment by granting the ability to allow our children to cue in and follow their life's purpose, to expose them to OurStory as opposed to His-Story, to teach them to maintain their temple (body) of God according to their biological make-up and not others', to create an atmosphere where they're not bombarded with external distractions that do not line up with their goals/ambitions, to introduce and assist them through exposure and application how to tune into their higher self (through spirituality) as well as help to correlate the laws of nature by way of science and mathematics to their everyday living.

Being "Earth" had many meanings to me throughout the years... there was, Earth represents the bearer of life, a relevant aspect in Gods Universe, a body who depended on the Sun for its sustenance and ability to see energy transform darkness into light...from a thought impulse created in the mind to the formation of matter. I came to recognize being an Earth as an honor, because she was the personification of God's intelligence which meant she would emit the radiation of her life giver, her Sun. To yield good harvest, an Earth should gain the humility to recognize her position and responsibility and carry it out to the best of her ability so it may serve her personal relationships as well as be an example for those seeking a new way of righteous living. Knowing that, being Earth is being God's co-pilot, the 'womb of man' that brings life to this plane. She is the way ;-).

Coming into this culture women must understand: The teachings are not to be underplayed, but respected as one would an elder though you may not fully understand them; That throughout your journey in learning self, you will unfold much relevant history and clarity; That our mind will not comprehend the wisdom in these "simple lessons" when we are filled with blockages from our upbringing or present circumstances. We must learn to take control of negative emotions or feelings brought on by staggering low vibration frequencies such as fear, doubt, anger, hatred, resentment, victimization, etc., which contribute to ignorance, disconnect, and remaining detached to the blessings of 120 Degree lessons. Key characteristics contributing to the understanding of the embedded wisdom within the culture are:

- *Identifying your foundational influences to any feelings/ ideologies/customs.*
- *Humility*
- *Patience*
- *Purity*

- *Respect*
- *Hunger*
- *Dedication*
- *Determination*
- *Proactive Health Practices*
- *Loving Attitude*

This way of Life is truly the best thing that could've happened to me. Being a part of the 5% Nation of Gods & Earths has been a key and an essential tool in my growth and development of self as a woman and an Earth.

PEACE!
E-Queen Glorious Earth, Now Justice (New Jersey)

There are many women that need to hear and read these words from caring sisters. Some men lack the requisite approach to elevate the minds of broken women and girls. Some men are even responsible for breaking the minds and hearts of women and girls, so a strong or wise sister is needed to rebuild the weak or unwise sister. Women communicate and relate to each other in ways that are different than how men and women communicate. Girls' friendships focus on making connections — which is essential to this process. Sharing secrets, relating experiences, revealing problems and discussing options are essential during girls' development. This differentiation in youth leads to dissimilar communication styles in adulthood. Women communicate through dialogue, discussing emotions, choices and problems. Men are usually action-oriented — the goal of communication is to achieve something. Sometimes, sisters need to figure things out on their own and just need the right person to show them the way explains Queen-Math3matiqs (pronounced "Mathematics"):

Peace! Prior to 1998 I was not inclined toward religious beliefs so for all intents and purposes I consider myself to be an Atheist. I did not believe in God, at least not the God that approximately 2.2 billion people or 31.5% of the world worship today. No one could explain God to me that made sense and aligned with my own logic and reason. It was very difficult for me to fathom that all things tangible were controlled by something so intangible. Fortunately it was their KNOWLEDGE of God, WISDOM of God and UNDERSTANDING of God that was incorrect. The Blackman is GOD, therefore God is real. Confidently I can say nowadays I bear witness that God exists. I pay homage to the teachings of the Nation of Gods and Earths that took the time to guide, support and encourage me to continuously gain the knowledge of myself. There is no need to rely on faith when you know.

Being a new born in the Nation was probably one of the most essential life changing experiences in my entire existence. For the first time some of my most pertinent questions were either answered or more importantly I was provided with a fool proof formula to obtain the answers on my own. This formula is simply to read not only read but study and understand with a black awareness. The formula is never relying on one source of information but having a multitude of resources. The formula is to ask questions, to question everything and take nothing on face value. The formula is to thine own self be true meaning to tap into that inner being not allowing yourself to be manipulated by outside forces that is going against your true self. The moment I decided to stand on my square or elevate myself I began to transform. My transformation was a direct result of my mental fortitude. My mental fortitude is a direct result of KNOWING, not guessing and with that I had the confidence to be a Queen by any means. I learned the significance of being and representing that in all aspects of my life. The significance is to know your worth. I learned sovereignty; mentally and physically.

Over the years my CULTURE nurtured, supported and reinforced the importance of self-preservation, the importance of family and the importance of solidifying my community. It is the information that I have obtained during my journey that allowed me to release the feeling of abandonment, not truly identify with a tribe or more specific ethnic group. Now I embrace or relish in knowing I am part of the whole and do not feel the need to identify with just one. I am more POWERful now than I have ever been because I live my life according to sound reasoning and sound reasoning exists within MATHEMATICS. Mathematics is and always will be the catalyst to greater things. It was and always will be the foundation that propels me to higher heights, greater views and broader perspectives. I am Earth. I am Queen Math3matiqs.

As we learn about ourselves and our history, we've come to learn about the damaging effects of miseducation on both men and women. Miseducation does not only take place in school, but also through a person's given religion. It is put this way because most people are given a religion or are made to feel compelled to be religious. They rarely choose to be religious. We do not see this as freedom. As a culture free from religion, we learn about God without globally accepted theological dogmas. It is also a challenge to remain a culture without people mistaking the culture for another religion. Beautiful Umi Earth explains further:

Peace! My name is Beautiful Umi Earth and on the date of this writing, today's Supreme Mathematics is "Build or Destroy". We use a different principle to build on, discuss and apply to our lives on a daily basis based on or in concordant with the day's date for each month of the calendar year. "Build or Destroy" is the 8th principle of our Supreme Mathematics and today happens to be the 8th day of the month. With that being said, this is the very reason why I chose to become a part of the Nation of Gods and Earths; to build,

destroy and refine certain aspects of my life. In the fall of 2009 after leaving the Christian faith, finalizing a legal divorce and pilgrimaging through a few other schools of thought, I began my journey of intense research of the Nation of Gods and Earths. I no longer wanted to be affiliated with organized religion. It was simply no longer serving me. After meeting Gods and Earths in the Atlanta area, attending their nation functions, continuing to read about and research the history of the nation I decided to commit myself to the teachings and embrace the culture as a single woman on October 31st 2010.

Being that I am still here and active 4 years later there are plenty of experiences that I could mention to fill the gaps of those said years. However, I would like to elaborate on what it was for me that kept me here. I was able to observe the nation and see this is a place where the people are actually in tune with their biological supremacy as original people on the planet Earth. We are the genetic prototype of all mankind. Also, it was where I observed people being accountable and responsible for their own prosperity, peace and productivity. I loved the whole look of the women in the nation and thought it was so beautiful, refined and refreshing. The women are promoters of health, wellness, sisterhood and unity of the black family. Gods and Earths are sharp thinkers, intelligent communicators who use credible resources to show and prove their point. All of this is caused by individuals alone and not some entity in the sky. This is what I wanted to do. This is what I wanted to live. This is how I wanted to define and express myself. Build myself up mentally, emotionally, intellectually and holistically. And destroy the miseducation, misunderstandings, misguidedness, indoctrination and traditions that no longer were working for me and were stunting my growth. Once something is destroyed, its demise is irreversible. There was nothing left to do at that point but to renew, rebuild and restore my life. Being a 5%er along with being an active part of a Nation of supreme people has shown me what it really means to stand on your own two feet, understand your abilities and execute your plans to be successful in your own endeavors.

As the years, months and days passed and I became more consumed with the teachings, I took notice on how some people in this nation define God. Now in this nation we have a very famous saying that is used which is "the black man is God". According to how a God or Earth defines

God is what determines to me if they really understand the teachings. The fact that the nation was built on the premise that the black man is god or supreme or first, simply means that every other race of people is genetically inferior to the black man and woman. Not that the black man has some sort of supernatural powers, or some type of X-Men qualities that defies the natural order and function of human abilities. Or that he is a spook or mystery god behind the clouds that is dictating the actions of people here on Earth like a tyrant or puppet master. Now if there were another word in the English language that defines a supreme being; supreme meaning first and foremost, being meaning living thing, then we would use that word. Nonetheless, the English language defines in our Standard English dictionaries this is what god is and science shows that he IS the black man. Once I hear someone's views on how they see God that shows me if that person really has this thing right or if they have simply used this nation as a substitute for religion and have wasted time searching for that which does not exist. Like before I mentioned that I chose this nation after removing myself from religion. It was a very lengthy process to come back to my senses. Religion had me literally out of my natural mind. And now that I have done the work to renew my mind and think for myself I can easily detect when someone else is struggling in that very area just by how they define the word God as we use it. I have searched every part of my being and I am certain that I do not have one ounce of mystery god left in me. When someone who has been a part of this nation for 20 years talks about how "Allah is looking down on them" or how "Allah is their sustainer", or "how the hand of Allah is protecting them", or how "Allah is blessing them", I just don't see them as getting it and it makes me feel like they have wasted their precious days here believing in something or someone that does not exist. I have destroyed those myths about spiritual entities assisting with human tasks. This life is all about us and what we are capable of achieving before we physically expire. That is why we say we come into the "knowledge of self". Now for someone like me who worked so hard at reversing the indoctrination taught to me to keep me illiterate chooses a nation supposedly built on facts and not belief see these type of teachings here it becomes very disturbing to say the least. These are the kinds of teachings that need to be destroyed. If you are here and striving to build yourself up in order to be your ultimate YOU, make sure you understand what these teachings are truly about before you commit your life to them. Make sure you are not substituting this culture for a religion. Make sure you are here to learn yourself and how you fit into the grand scheme of things so that YOU can teach your children and THEY teach our community so that WE can teach the world. Peace!

I come in the righteous name Umi'Sudan Lunar Allat Wisdom (Umi'Sudan L.A.W.) reflecting the loving light of my maker & owner Lord Hashim M7 Allah. We reside in Divine. Allah — Truth.Known (Dallas, Texas). I am the Umi (mother) to 1 beautiful daughter (Nefertari) and nurture countless other children in my community. The 1st step of my journey began in 1997 and through many obstacles I remained focused; I call this period from 97' -01' the Purge. From 2001 to the present I have been included in published works like the Def Jam Poetry Anthology, "Bum Rush tha Page" and continue to produce creative works as, Black Page. "Traveling" a track recorded and released in 2012 on the. VORTEX project, will also be part of the NGE 50th Compilation CD set. Visit me online @ www.artistblackpage.wix.com/home –

In addition to being an artist I am also a business woman. So I created Brown Mood Consulting in early 2000 and now with my God manage a variety of projects. One project in particular that we are extremely proud about is our involvement with our nation's 50th Anniversary Celebration Concert (October 3rd-5th 2014 @ The Apollo Theater) "The God & Hip Hop Weekend". Since 2001, Sista Touch Natural Haircare Services has been another medium I have used to reach sisters and brothers in my community. Being able to address my clients' different areas of interests, from diet to gaining the "Knowledge of Self", is very rewarding. 2011 began a new Koran for me, my walk in 120 with my enlightener and beloved God has been enriching to my life. I build even harder, with a reaffirmed investment into self and cipher. I knowledge my lessons in 2012, and now each quarter I facilitate a conference call that began in Region 5. The call features Earths in the nation with valuable history and who are move at a terrific rate of speed. It is a great opportunity, for any original woman striving to gain the knowledge of self, to experience some of the beauty this nation has to offer. Connect with the NGE in Sudan @ www.ngeinsudan.ning.com (ngeinsudan@gmail.com) for details on the Worldwide Earth Conference Call Cipher (E.C.C.O). Queen Cipher said, "I come to the cipher to cee in my sisters what I come to show and give, and that's—EQUALITY!" When I heard her build on that point I thought of two words, "limitations" and "access". So let your axis be Supreme Mathematics & Supreme Alphabets to unlock and gain access to the limitless possibilities available to you, when you live the Righteous way—PEACE!

By now, you may have heard a lot mentioned about 120 Lessons. This shows how 120 Lessons plays such an important role in this culture. It is not just questions and answers on paper that should be memorized. For us, 120 Lessons is the red pill we choose to take to learn the truth of reality instead of the blue pill's blissful ignorance of illusion that's chosen by the masses of pop culture. Its truth is so potent some brothers have tried to limit their counterpart's access to it. Here we are a people in need of holistic medication, but some men have taught their women they only need some of the medication because of his insecurities or short-comings. They would say *"My Earth doesn't need to know 120 Lessons."* Now, who's putting a limitation on whom?

All the Lessons are meant for men, women, and children to study. In fact, to deny anyone knowledge in this regard is to be no better than a slavemaster who denies his slaves to read. I am proud to know so many Earths who were taught and raised properly by true and living Gods and Earths. Earth I Asia, for example, is one of many who were born and raised in this culture by her parents Allah's Justice and Meccasia. She has organized and hosted Earth Ciphers and Civilization Classes in Pennsylvania, New Jersey,

New York and Connecticut. Regarding the importance of an Earth knowing 120 Lessons, Earth I Asia wrote:

I once asked a close brother of mine, what does it mean to be an Earth? His reply was you have to have the magnetic of the moon with the qualities of the Earth; one of the strongest attributes of the earth is equality. Equality means to be equal in everything. Equality doesn't mean it has to be the same it just indicates that you gave what you can give. This needs to be shown through your ways and actions. Your ways and actions are the sum of your thoughts, ideas, and judgments. What you do should come from what you know. Over time what you do develops into your culture (how you live on a day to day basis). Being an Earth not only means you have knowledge of yourself, knowledge of who and what is god and knowledge of you role in the universe it also means you have a set culture. The culture that you live comes from the values and traditions that have been passed down through previous generations. 120 is one of those values that have been handed down to us because previous generations were wise enough to see the value in it. Now as a nation we have traditions associated with the study of 120.

Traditionally speaking in the Nation of Gods and Earths 120 is an important building block of your growth and development as you begin to gain the knowledge of yourself. 120 is essential because it teaches discipline, it shows dedication and is a rites of passage within this culture. Learning Supreme Mathematics, Supreme Alphabets, and 120 lessons' requires you to take time and energy to memorize, quote and understand it. If you claim to love this culture but don't want to participate in something that is at the very heart of this culture I would question what is the reason for this. Is it that you don't have the commitment level and discipline that it takes to go under that study and master the lessons, or do you just not see the worth in learning your lessons therefore you don't apply the effort. Whichever or whatever the reason you are denying yourself vital tools that will help you build a strong foundation. What can I say to that other then you need to ask yourself what are you here for? The true and living Earths and Gods will be asking you the same. We are a nation of who prides ourselves on being able to show and prove. We have accomplished a lot in short history of this nation and we still have a lot that needs to be accomplished. Think about this for a moment when you have a recipe for a dish you want to cook but you are missing a key ingredient. Do you make it anyway or do you wait until you are able to complete your ingredient list. If you make it anyway as the chef you will always be able to taste that something is missing it just won't be the same. Even if you are able to cover up or try and compensate for the missing ingredient you will know it isn't there. When you come together as a nation we speak our language supreme mathematic, supreme alphabets and we apply these two to our understanding of 120. How fluent do you want to be in this language and this culture? I for one want to be able to speak to my sisters within this culture without feeling obligated to be politically correct and not use certain language or degrees to try to accommodate those who simply choose not to study. Those who did their duty shouldn't be penalized for those that didn't. The more knowledge you have the more power you have to make positive change. We need our ranks strong and tight we don't need any weak links.

Many will come few will be chosen. The chosen few have to choose themselves. You have to show and prove that you are the said person with the ability to accomplish the said task. You can't claim to be a surgeon without the skill set to operate on someone and you shouldn't claim to be a part of a culture if you are not willing to live within the guidelines of that culture. To become the person with that said skill set it takes time and devotion. You can't become a master with minimal study just like you don't become earth overnight. You have to put in that work because it all comes down to show and prove. You have to prove it to yourself first. We as Earths make the conscious choice to walk the righteous path. We make our knowledge manifest to wisdom. We value understanding above all else. We are able to apply this knowledge, wisdom and understanding to our everyday lives making it a part of our culture. This allows us to refine our ways so we aren't living like savages in the pursuit of happiness. Are you ready to choose yourself? In order to be successful in life you need discipline. Discipline is a warrior's best friend and a coward's worst enemy. Show and prove to yourself and your nation that you have what it take to undergo one of the foundational studies of 120. Peace.

HOW TO SET UP EARTH CIPHERS

BY SCI-HONOR DEVOTION

Ultimately, sisters come to ask about Earth Ciphers. What is an Earth Cipher? How do they start? What is the structure of an Earth Cipher? What topics are talked about in an Earth Cipher? These questions are best answered by Earths who've learned the culture and tradition from Earths in the preceding generation. Traditionally, the women in this culture set up sister circles where they can form offline bonds by going over lessons, express their concerns, and exchange ideas. To ensure meetings stay in the right course, it's beneficial to know how to set up a structure for Earth Ciphers. SciHonor Devotion shows how an Earth Cipher can be started with having specific agendas geared towards the education and empowerment of sisters:

> **The Earth Cipher's Determined Idea — An Earth Cipher is a collective meeting of women who have committed to living the 5% culture, yet can also include women who are interested in the teachings of the nation, as well as the culture and conduct of The Earths within. Typically the host of the Earth Cipher is well versed and actively experienced within the nation. If she is not, she should be guided and assisted by an Earth who is qualified herself to host a cipher. This allows the new host to gain experience, share her knowledge, build confidence and also ensures that the teachings do not become mixed, diluted and tampered with. Earth Ciphers can cover a myriad of topics ranging from community building, diet and nutrition, sisterhood, financial stability, cultural etiquette, sewing, how to keep our Gods, pregnancy, breastfeeding, reproductive health and much more. The Earth Cipher brings together women who subscribe to the teachings of Allah and is open to women of all conscious paths allowing us to build bonds with each other and our communities.**

Below is a Sample Earth Cipher Agenda. This format has been implemented in various regions throughout the nation. *<u>Sample Earth Cipher Agenda</u>*

November 17th, 2015 / 12:00pm – 5:00pm

Agenda Topic: Who is the Earth?

Defining the Role of the women in our nation. What is her responsibility to herself, her God, her Family, her Nation...? Today being the Queen Day (the 17th). What are the Characteristics of a Queen? How do we hold each other accountable when coming in the name?

Introductions: *Who are you? Where are you from? Who is your Educator or God? Have you knowledged 120? What degree are you on? What do you look to get out of attending Earth Ciphers? What are some skills that you may have and be able to add on to future Earth Ciphers? Do you have babies and what are their ages? *Write your name on a sheet of paper and put it in the bag. (To be picked during the expression period.)*

Let's Build: *How do you see Today's' Supreme Mathematics? You can add on with wherever you are in your lessons.* **120 Check-In:** *If you have not knowledged 120, you should expect to be asked if you are any further in your lessons than you were when the last Earth Cipher occurred. Be prepared to build on the degree and your progress.*

Earth Science and Beyond: *Share a fact with the cipher that you have researched or even stumbled upon in relation to Earth Science or any field of Science that may interest you or the cipher.*

Healthy, Strong and Good: *In this section, we keep in mind that the Earths have always taught the three P's—PREVENTION, PROTECTION, and PROMOTION. We use the NGE Monthly Health Calendar as a guide and to be able to study the same things our sister Earths (and Gods) are studying in their ciphers. So, In accordance with our NGE health calendar the Herb of the Month is Parsley. (We strive to announce what next month's focus and herb will be so that studying/preparation can be done. Next month, December 2013 is Frankincense Oil) –*

NOVEMBER TEACHING: Kidney Stones: HERB OF THE MONTH: Parsley

Parsley is extremely rich in Vitamins A and C. It also reduces inflammation and cleanses the body of toxins. Parsley has been used traditionally to treat urinary tract infections, kidney stones, and liver, bladder and prostrate problems. NOTE: Parsley oil, juice, and seeds should be avoided during pregnancy because they are uterine stimulants. Parsley eaten in food is fine, but pregnant & nursing mothers should refrain from using parsley medicinally. Kidney Stones—A kidney stone is a solid piece of material that forms in the kidney from substances in the urine. Kidney Stones range in size from as small as a grain of sand or as large as a pearl. Most kidney stones pass out of the body on their own. Sometimes stones get stuck in the urinary tract, blocking the flow of urine and can cause lots of pain. If you are experiencing any of the following symptoms, you need to get professional assistance dealing with this. Contact BeNature Be Earth or Divine Iz Earth for advice on how to proceed. • Extreme pain in your back or side that will not go away • Blood in your urine • Fever and chills • Vomiting • Urine that smells bad or looks cloudy • A burning feeling when you urinate

"Anyone with problems or health conditions related to any of these herbs or their actions should not use them. Some herbs are known to react with your medication. Please consult your physician or NGE health practitioner before starting on any herb." NGE Health Calendar Committee

**If you are on Facebook, you may be interested in joining the NGE Health, Fitness and Martial Arts Group.*

Sew, Cook and in General How-To...: *We will do a Raw/Live Food Prep Session in which we will include parsley in the recipe. We are the Earths. We do not promote artificial foods but foods that are beneficial for our bodies so that we become and remain healthy, strong and good. Some ciphers allow meat to be prepared and served while others do not. If you are attending, and it has not been specified, you should contact your host to find out what they allow. If the cipher is in someone's home, they may not allow you to bring meat or other artificial foods. The NGE overall does not deal with any pork, pork products or products with pork in them. We also do not consume any form of shellfish or tuna. Some people do not eat collard greens and kale while many vegans do because* *of their nutritional benefits. Again, ask if you are unsure. Overall, we are striving to instill and practice good health and eating habits in all of our ciphers.*

Theme: *Defining the role of the women in our nation — We will be discussing some history as it relates to the women in this nation through personal accounts, written accounts and oral history using a few "14th Degree and Beyond" Magazine interviews and recordings to refer to. What stories have you heard about the Earths who helped to lay the foundation for us? What is the importance of some of the standards that the Earths set for themselves and how can we continue to set standards for ourselves as Earths today?*

Expression: *In this section of the cipher, we would ordinarily create something. In relation to today's mathematics, our build today and today being the Queen Day, pick a name. Tell the person whose name you chose, something that you see in them that is "Queen"ly / "Earth"ly, something that you think you could learn from them or even ask them something that you would like to learn about them. Other options could be to sew or paint something as it relates to being a Queen.*

Prepare Ourselves: *In this section, we plan for next month. We set a date, location and agenda topic. Whoever hosts the Earth Cipher will create the agenda and is responsible for letting the attendees know what to expect, prepare, or bring.*

Homework:

To Present: *Summarize what you learned during today's Earth Cipher to share at the cipher next month. What stood out to you the most?*

Earth Science and Beyond: *Please bring your Earth/Science Fact to share with the cipher.*

Healthy, Strong and Good: *Please strive to learn something about Frankincense so that when we meet again, we can discuss it. It is December 2013's Herb of the Month*

Self-Reflection: (doesn't have to be shared unless you would like to) What things about yourself do you think that you need to improve upon in order to live up to your full potential as the Earth that you aspire to become?

What to bring next time?:

-1 — composition notebook/tablet, binder, or other note taking tool for Earth Ciphers, Civilization Classes, Rallies and Assignments -A snack to share –Suggestions of things you can bring — fresh fruits / vegetables (cleaned and bagged prior to bringing them if no sink will be available) apples, bunch of bananas, bunch of grapes, etc.), plain chips / tortillas, hummus, guacamole, 100% juice, dried fruit.

When planning an Earth Cipher agenda, we may even plan the whole agenda topic around what the herb of the month is during that particular month. Ex: If the herb of the month is Red Raspberry, the agenda may be "Women's Reproductive Health". In that case, we could study our physical bodies in addition to other topics such as Eugenics, have book reviews, like Killing the Black Body or Medical Apartheid. Our physical activity may be something like Belly Dancing which requires a lot of movement in the womb area. We may also prepare herbal teas that could benefit us during menstruation, pregnancy, or just for our reproductive health in general which would include the Red Raspberry herb. We may watch some videos dealing with reproductive health or menstrual care and create Menstrual charts or journals to track and record our cycles. And we may do some product review discussing products on the market that deal with women's reproductive issues.

So you have an idea of how to create an agenda that will show Earths in motion. This is the format that I created for use at our C-Truth Earth Ciphers. This format can be adjusted to your liking and customized to the needs of the cipher in your area. You can destroy and build on what you found useless or useful to you in this sample.

Some other Earth Cipher Agenda ideas: — Building Financial Wealth — How to Raise our Daughters — How to Raise our Sons – How to Deal with Effective Communication – How to Handle our Emotions as Women – How to Choose a God / Educator – Public Speaking – Living off of the Land – Refinement and 3/4ths – How to Start a Home-Based Business – Survival Skills for Women – Self Defense – Sisterhood – Conflict Resolution – Critical Thinking and Using Logic/

Using Supreme Mathematics – How to Act at Home and Abroad – Dealing with Issues in the Workplace – How to Appropriately Interact with 85% — Dietary Science – Self Esteem – Our Nations Foundation "The Early Years" — and More.

Peace,

Hosted by Sci-Honor Devotion

CHAPTER 6

Response-Ability

RESPONSIBILITY AND ACCOUNTABILITY apply to ALL PEOPLE WHO SAY THEY ARE RIGHTEOUS. What does this mean? When it comes to being a part of the Five Percent, whether you have been here for one year or fifty years, you have to clean yourself and how you live up if you want people to take you seriously. Nowhere can it be said that these teachings promote sitting around, hanging around and taking up space. Whether we realize it or not, we are all supposed to be building. There is no build too large or too small for this nation, but we have to build. All the guesswork can be taken out of what we truly teach when we are

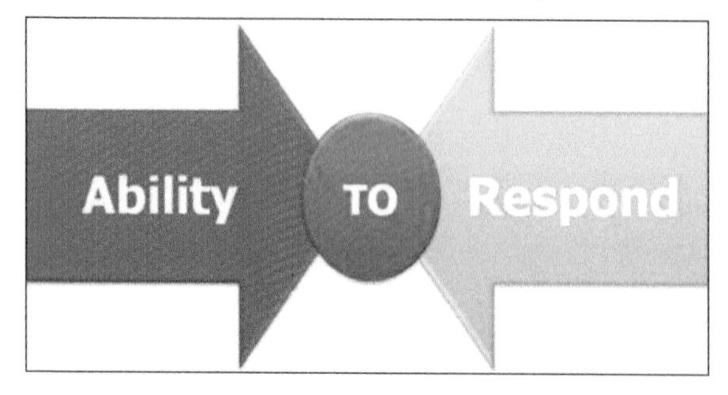

responsible and accountable. I confess, as in any group of people, there are those who are positive examples of what we teach and those who are *just here* with the same old thinking and ways of the eighty-five percent. I used to be the same old knucklehead until I became responsible for my own thinking and behavior. People started to look and respond to me differently when I acted in accord with what I knew to be the truth.

One of the essentials of leadership is to seek responsibility and take responsibility for your actions. Defined, this means to take the initiative in the absence of someone having to tell you that you ought to do something and accept the responsibility for your actions. Being responsible means stepping up to the plate to perform your duty. Some have mozied into this nation thinking all they have to do is quote lessons and that somehow makes them true and living. Some stopped coming around simply because they were afraid they would be asked a lesson or because they didn't have the discipline to change their diet. Your duty isn't just about quoting lessons to the next person. This is one of the reasons why few people are left to tackle the hard tasks while others just want to claim glory and victory without putting in any work. This includes everything from taking care of home, raising your children, and bringing youth to Parliaments to having a role in nation activities, being involved in some community work, and advocacy (if you qualify yourself). Let's not forget we are nation building and that requires small and large work on everybody's part.

Just as a person shouldn't claim to be a champion if they hadn't had a single fight, people shouldn't claim glory if they haven't done anything personally to earn that glory. Real Five Percenters don't jump on bandwagons and the nation shouldn't be seen as a bandwagon for everyone to jump on. Another aspect of being an actual nation member is the wearing and representation of our Universal Flag. Whether people know it or not, there is a degree of accountability that comes with wearing this flag. This accountability is not limited to being able to know your lessons or the history and meaning of the flag. That's only a part of it. For some, that's all you need to know to get a pass.

Back in the days, a person had to be someone in good standing to wear the flag. You couldn't be involved in crime, not attending school, or being a social misfit and think it is okay to walk around with the flag on. You could be stopped at any given time by a God or Earth and be asked a series of questions because it was not something that just anyone would wear. Some didn't survive those questions and flags had to be removed or confiscated because the person wasn't right and exact. For some on a small level, it was about who can take the most flags. But for others, it was more about maintaining integrity and honor.

We have seen people selling merchandise with our flag on it and people wearing the flag without a clue as to what it signifies or what responsibilities come with it. Many of these merchants and false flaggers are fugazie. In our world, it not only symbolizes the natural unification of the Black family, it depicts our ideology and commitment to the culture. It symbolizes universal (or natural) law being lived out by the Black man with knowledge of self who is functioning at his highest level of intelligence and is doing his duty as God, the Black woman fully refined, in-tune with her culture and bears witness to Allah's Supreme Mathematics, and the Black child with infinite potential to rise to the top. Therefore, if you are not ready to live the life, you should not be wearing the regalia. This includes crowns worn by the men and three-fourths (geles) and head-wraps worn by the women.

Commensurate with our cultural motifs are the expectations that your word is bond and you are living to exemplify the righteousness you advocate. Imitators are always a poor example of who we are. They give us a bad name and many people out there have been mis-lead by *Jive Pretenders* (a suspect Five Percenter). People in the community know all about those who used to "preach" and not "teach." Whether they knowingly failed to properly teach someone or knowingly did something out of character, a dubious Jive Pretender is more like the 10% (in my opinion) in that their actions further concealed who we really are to those who are most in need of knowing who we are.

Another shady type of irresponsible person is the gossiper, always having a problem with people but never having the guts to address the problem in a face to face encounter. Instead, they choose to talk about them behind their backs. The gossiper that spreads gossip is just as bad as the gossiper that first speaks it. Gossip is idle talk or false rumors about the personal or private affairs of others. Gossiping in this sense not only reflects weakness, it reflects wickedness. It is usually done to drag a person's name and reputation through the mud. It is a weak and wicked act that should have no place in our way of life.

> *"Do not concern yourself with things about which you have no knowledge. Verily, your hearing, sight, and heart — all of them will be called to account"*
>
> (QURAN 17:36).

> *"Oh you who believe! If a wicked person comes to you with any news, ascertain the truth, lest you harm people unwittingly, and afterwards become full of repentance for what you have done."*
>
> (QURAN 49:6).

"Oh you who believe," (meaning ***you who take stories on face value***), the spreading of misinformation about people can make a nation weak. Today, we can't afford to be morally weak because so much (including the lives of our children) depends on our unity and strength. How can a nation sustain unity and strength and people talk behind each other's backs all day instead of building? So much hate, resentment, jealousy, distrust and fighting is caused by the participators of gossip. We have to be responsible enough to stop gossip at its root and not allow it to live by not starting it or spreading it.

Knowledge of Self calls upon us to bring out the best in ourselves and in others. How does anyone with Knowledge of Self or seeking to get Knowledge of Self have time for that? We should validate our sources, and not engage in conjecture and taking things on face value. We want to build a nation, not destroy it before it even gets started. If you have been the subject of gossip, here are five tips to handle the situation with aplomb:

1. **Rise above the gossip**
2. **Understand what causes or fuels the gossip**
3. **Avoid spreading gossip**
4. **Allow the gossip to go away on its own**
5. **If it persists, gather facts and witnesses and try to clear your name by confronting the gossiper(s) in a civilized manner.**

Response-ability is all about our power to properly respond (*not react*) to evil or wickedness in its many forms. The more of us respond adequately and appropriately, the better the outcome and another victory for home team. With knowledge comes great responsibility. Therefore, for each person that claims to have knowledge, there is a responsibility that comes with that like teaching your children, your mate, your neighbor and/or nine young people, by living this culture out at home and abroad (this includes social media).

On the topic of ***Building***, it shouldn't be limited to or colloquialized with talking. Some people think the longer they talk, the more they're building. This term was not intended to mean have endless discussions, debates, arguments, rants, etc. It actually means to add something or make something. Those new to these teachings should get this misconception right. If we are truly nation building, a nation can't be built on quoting lessons for years, months, and days and not going to work. If we could agree that we won't argue over how we see a lesson for just one year and instead organize ourselves to acquire property and facilities, we can make a world we can call our own. There's much that needs to be made in a nation and we need the responsible Makers and the Owners (*not Talkers*) to start making some things happen.

There's also a responsibility to ***Destroying.*** Coming up in the teachings, we often would say we "destroy negativity within our cipher." We learned that energy is conserved over time, therefore, it can't be created nor destroyed (energy exists in potential and kinetic forms). Energy can change form and be transferred, but it can't be created or destroyed. Instead, what we must actually destroy are things that waste our time and energy. How do we do this? By changing how we use our time and energy. When we first got lessons, the knowledge entering our young minds helped us to see reality more vividly. For every second, minute, hour, and day we studied we were able to transfer or change the direction of our energy from negative to positive and from wasting time to using time wisely. The more knowledge we acquired the more responsible we became to destroy anything that impeded our growth and development. We had to Build and/or Destroy to be Born complete.

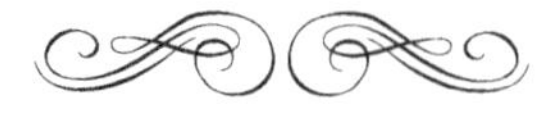

"Build a nation or be destroyed."

—ALLAH

With the advent of social media, many have been sucked into ceaseless online debates. Most of these debates are not even worth our time and energy in light of the duty Five Percenters have to their family, community, and young aspiring students. Oftentimes, the people engaged in such arguments have never actually met and have no clue the true identity of the person on the other end of the computer. Some of these *"Internet Gods & Earths"* have no students and are not teaching or living this culture at all. They don't show up to any nation events but have hundreds of Gods and Earths as Facebook friends. The nation comprises of real life people you seek out and find. The truth we teach and exemplify is seen and heard not copied and pasted.

We are all here now seeking solutions. Our mission continues to be to teach, elevate, and improve ourselves. Therefore, any serious person who is or wants to be a Five Percenter, a righteous person, a conscious person, or a revolutionary person, has a responsibility to destroy (remove or take away) anything in your way of achieving your goals. Here are some things we have the response-ABILITY (POWER) to destroy:

Envy (When we lack a desired attribute or success enjoyed by another person).

Jealousy (When something we already possess is threatened by another person).

Hatred (When we passionately dislike ourselves or others because we lack understanding of ourselves or others).

Greed (When we uncontrollably desire to live beyond our means to possess wealth, goods, or objects, far beyond the needs of basic survival and comfort).

Wrath (When we think or feel our boundaries have been violated and we are filled with uncontrollable and retaliatory anger towards someone who could be totally unaware they've crossed a boundary).

Sloth (When we simply don't care about our well-being, our duty, our lessons, our education, our youth, our schools, or future; we avoid the mental and physical work toward our cultural obligations).

Pride (When we irrationally and unreasonably *"Ease – God – Out"* from a false sense of self-aggrandizement or status; the complete lack of humility; the *"I'm God-no-one-ca n-tell-me-nothing-syndrome"*).

Lust (Similar to Greed; when we (men and women) have an *insatiable* need to have sex, eat, or some other activity).

Gluttony (When we live to eat instead of eating to live, we over-indulge in foods and drinks that make us overweight and unhealthy).

This 50-year point and beyond, marks a certain level of growth and development in this culture. If we do what we always did, we will get what we always got. The teenagers of yesterday are grown with babies of their own today. We have matured in the way we see the lessons, the way we dress, the way we eat, and the way we live. This is the elevation real Five Percenters are dealing with which is why so many in the community don't see us on the street corners any more. There is a new generation of young people on those street corners and most of them have no Knowledge of Self whatsoever.

It's our duty as Five Percenters to give good orderly direction to people younger than ourselves because they are the ones who will carry it on. If you have Knowledge of Self, if you're conscious, righteous, civilized, or a revolutionary, here are some considerations that can make a difference in someone's life and in the community:

1. **Interact with people offline as you would when you're online. There's only one reality.**

2. **There are two ways you can be wise: a) Know everything about something; b) Know something about everything.**

3. **Always choose the best part (Make the best of choices in all that you think, feel, and do).**

4. **Preserve and pass down the 4 parts of our culture (Language, Literature, Art, and Music) to your children.**

5. **Five fingers together make a mighty fist symbolizing the strength of our unity. It's time to use our power to fight back.**

6. Stop being afraid of what the enemy might think of what you say and do. HE IS LIMITED AND SEEKS TO KEEP US LIMITED TO HIS RULES SO HE CAN MASTER US. The fear was planted a long time ago. ELEVATE ONE STEP BEYOND those limitations.

7. Be who you say you are! This applies to men and women! WHO YOU SAY YOU ARE SHOULD BE A STEP BEYOND THE LIMITATION YOUR ENEMY PLACES ON YOU.

8. Build and waste no time waiting on people to do for you what you can do for yourself.

9. teach nine others what you learned.

CHAPTER 7

God Grabs the Microphone

"I'm the intelligent wise/on the mic I will rise/
Right in front of your eyes/cause I am a surprise/
So I'm a let my knowledge be born to a perfection/
All praises due to Allah and that's a blessing/
With knowledge of self, there's nothing I can't solve/
At 360 degrees I revolve/
This is actual fact, it's not an act, it's been proven
*Indeed and I proceed to make the crowd keep moving."**

IN THE LAST fifty years, we have seen more and more rap artists proud to infuse being a god in their lyrics. Artists who are knowledgeable about being a god in their upbringing were either exposed to the teachings of the Five Percent or were (some of whom still are) Five Percenters their selves. The spoken word is an art form of self-expression that Five

* Rakim Allah, *"Move the Crowd,"* off the *"Paid in Full"* Album

Percenters (namely Gods & Earths) woke the mentally dead by raising the consciousness of listeners with teachings of Knowledge of Self. A list of MCs was outlined in the first edition of *The Righteous Way, Chapter 1: The Golden Age.*

In the world of pre-commercialized, corporate poisoned Hip Hop, people emceed about reality as they saw it and lived it. It was raw and pure talent from the creativity of Original people living in the ghetto. Hip Hop music evolved as an outlet in the low-economic areas of New York City. It reflected the social, economic and political realities we lived in. Self-styled expression and originality caused Hip Hop culture to naturally emerge from our Jazz, Blues, Folk, Caribbean, and Negro-spiritual past. Our emceeing, deejaying, breaking and Graffiti writing made up the self-expressed elements of the culture. But, these elements were not the only ones sampled to give birth to the Hip Hop culture. There was another element that was the heartbeat that kept the culture going. It was, in my opinion, the foundation of the culture. It's what elevated the culture. It's what some consider the forgotten element — consciousness.

The DJ had to be conscious of what music to play or blend to get the crowd moving. Before an MC can move the crowd with his or her rhymes, they had to be in touch with the crowd. They had to know the crowd, where the crowd came from, what was popular at the time, and most importantly, they had to be conscious of what to say to move their bodies. The MC had to speak to the audience, entertain the people, and in general keep the party rocking. We had to draw on our knowledge and history and that required consciousness. It is hard to imagine a non-conscious person doing this with style and finesse. My Hip Hop education began with Rakim, Big Daddy Kane, Dougie

Fresh, Slick Rick, and Run DMC; but I soon learned about the originators: DJ Kool Herc, GrandMaster Flash and the Furious Five, Lovebug Starski, Afrika Bambaataa, and the Sugar Hill Gang. One thing they all had in common is that they were from neighborhoods where Five Percenters had a serious mental and cultural impact on the people.

According to Steve Greenberg, founder and CEO of S-Curve Records, disco was the last mass popular music movement that was driven by the baby boom generation.* Disco music was a worldwide phenomenon, but its popularity declined in the United States in the late 1970s because many Rock music fans did not like the culture-shock. On July 12, 1979, an anti-disco protest in Chicago called *"Disco Demolition Night"* had shown to be how the Rock culture would wipe out disco and its culture. In the subsequent months and years, many musical acts associated with disco struggled to get airplay on the radio.

Hollywood promoted violence, racism and sexual ilk. The porn industry promoted the exploitation of women and the devil worshipping Metal bands all enjoyed mainstream commercial success. There was no Hip Hop or Rap music to blame then. In New York City, Hip Hop as a music and culture formed during the 1970s when block parties were rocking with DJs. Urban griots grabbed the microphone to tell their story and they grooved with the beats. During this time, large enclaves of the Five Percent could be found in the roughest neighborhoods throughout New York City. The teachings of Knowledge of Self spread throughout entire housing projects, street corners, schools, hang-outs, including the suburbs.

---

* Greenberg, Steve, *From Comiskey Park to Thriller: The Effect of "Disco Sucks" on Pop*, July 10, 2009

Of all the elements of Hip Hop culture, rapping became the most lucrative which opened the door for the world to begin studying and mimicking the entire culture, including droppin' science. Alas, the world will come to know about the Gods who laid down its foundations, fortified its walls and erected its pillars. It has become the number one influential music genre in the world. It became a force so profound and lucrative it would not suffer the fate that disco did by America's Rock culture.

> ***Rap is the rock 'n' roll of the day. Rock 'n' roll was about attitude, rebellion, a big beat, sex and, sometimes, social comment. If that's what you're looking for now, you're going to find it here.***
>
> —Simpson, Janice C.*

A vital point was the influence of Afrika Bambaataa and the Universal Zulu Nation. It should be noted, the birth of Hip Hop was in 1974, but Allah's teachings had been in many neighborhoods throughout New York City for ten years by this time. Bambaataa's influence in the Bronx River area as a DJ coupled with the influence of the Gods and Earths, were pivotal points in Hip Hop. Hip Hop listeners would be in for a new sound as the Gods began to bless the microphone over rugged beats. The teachings now had a medium to voice God-consciousness and Knowledge of Self which ushered in Hip Hop's Golden Era. Dasun Allah, in a blog entitled, *"The GODS Of Hip Hop: A Reflection On The Five Percenter Influence On Rap Music & Culture,"* aptly stated on hiphopwired.com:

* February 5, 1990). "Music: Yo! Rap Gets on the Map". *Time* (New York: Time Inc.) June 7, 2012

This was more than just music, more than just message. It was a manifestation of a way of life that is at the core of what we now know as Hip Hop culture.

Pulsating with Five Percenter vibrations, these lyrical alchemists metamorphosed the base elements of the teachings of the Nation of Gods and Earths into the golden wisdom of their songwriting. Indeed, some of the best to have ever touched the microphone have been students of, influenced by, or have utilized elements of the teachings of Nation of Gods and Earths, commonly known as Five Percenters.

*This includes Jay-Z, Nas, Rakim, Busta Rhymes, Wu-Tang Clan, Brand Nubian, Poor Righteous Teachers, Gang Starr, Big Daddy Kane, LL Cool J, Big Pun; even Erykah Badu and the Digable Planets whose Grammy-Award winning 'Rebirth of Slick (Cool Like Dat)" contains a line alluding to the Five Percent.**

When Gods began grabbing the microphone, their self-styled wisdom brought a new level of magnetism and energy to listeners. Fans of this new Five Percent perspective grew in consciousness and wanted more of the truth transmitted by God-MCs. Dasun Allah further added the supportive works of Professor Felicia M. Miyakawa and Russell Simmons to list of people who knew that the Gods were a dominant force:

In her study, Five Percenter Rap: God Hop's Music, Message and Black Muslim Message, documenting that Kool Herc reported a heavy Five Percenter presence at his parties, Professor Felicia M. Miyakawa noted:

"Even in the earliest days of Hip Hop, the Five Percenters were regarded as an integral part of the Hip Hop scene." The fact that an entire book has been published on the topic of Five Percenter influence on rap is a testimony to the strength of the impact.

* http://hiphopwired.com/2010/03/24/the-gods-of-hip-hop-a-reflection-on-the-five-percenter-influence-on-rap-music-culture

It has also been reported that two of Hip Hop's founders Kool Herc and Afrika Bambaataa personally studied Five Percent teachings as well. The foundational lessons of the Zulu Nation, the spiritual core of early Hip Hop, were directly derived from the Supreme Mathematics and Alphabets. The Gods and Earths being a factor in the switch from street wars to street jams gels with what Russell Simmons further records in his autobiography.

*"During the period when the gangs I hung with in the 70's gave way to 80's Hip Hop culture," writes Simmons. "It was the street language, style and consciousness of the Five Percent Nation that served as a bridge."**

Along with the Gods in Hip Hop included a list of notable female MCs who represented the culture in their lyrics. One female in particular, didn't just spew Five Percenter-influenced lyrics, she dressed the part quite well in stylish headwraps and 3/4ths of clothing which is the traditional dress code of the women in the nation. Erykah Badu was introduced to the culture while in college. In her youth, she had decided to change the spelling of her first name from Erica to Erykah, as she believed her original name was a "slave name." It was quite common for Five Percenters to be in colleges and universities as this was the educational direction Allah intended for them from the beginning.

* http://hiphopwired.com/2010/03/24/the-gods-of-hip-hop-a-reflection-on-the-five-percenter-influence-on-rap-music-culture

> *"As Erykah Badu, it has nothing to do with me, the way I look, my hair wrap, my style, it's about you and what you feel for my music. If I can make you feel like the way that people who influenced me made me feel, that's completion."*
>
> —Erykah Badu

Hip Hop pioneer and co-founder of Def Jam Records, Russell Simmons, reflected back on his own history and experience with Five Percenters in his latest book, *"Super Rich."* Simmons grew up in Hollis, Queens, at a time when God and Earth culture was widespread throughout the borough. In this excerpt, Russell Simmons describes the influence Five Percent culture had on Hip Hop, as well as on one of the most notable rappers of our day, Jay-Z, who has worn the Universal Flag instead of an iced-out cross or Jesus piece:

> *"Growing up in Hollis, I always thought the flyest cats in the neighborhood were the Five Percenters. For those who aren't familiar, the Five Percenters are an offshoot of The Nation of Islam that are named after their belief that people are divided into three groups: the eighty-five percent who are blind to God and the truth; another ten percent) made up mainly of elites like politicians , CEOs and members of the media) who know the truth but use it to exploit and deceive; and the final five percent, from which the group takes its name, who know the truth (namely that God is a Black man from Asia), but rather than abuse it, try to use it to uplift people.*
>
> *While, as you might imagine, the Five Percenters have never found much mainstream acceptance, they were very well-known in the hood for the flamboyant language and accompanying philosophy they created called "Supreme Mathematics." Growing up, I used to love to watch the "Gods and Earths" (as members of the group are known) stand on the corners and "drop jewels,"*

their term for philosophizing about religion and the true role of black folk in America.

While the Five Percenters' silky smooth style of speech would go on to have a significant, if underappreciated, impact on the language of Hip Hop, I personally never got too deep into their entire scene. While I loved listening to those smooth niggas hold court, ultimately I was more concerned with gang banging, getting high, or chasing girls to spend too much time thinking about the jewels they were dropping.

Recently, however, I was reminded of just how deep the Five Percenters really were after I fell in love with a song called "Exhibit C" from the rapper Jay Electronica. In the song, Jay tells a story of living on the streets ("without a single slice of pizza to my name"), where he wastes years "shootin' dice, fighting and smoking weed on the corners/ looking for the meaning of life inside a Corona." He remains stuck in that negative cycle until he's approached by several Five Percenters, who, as he puts it, "inform" him of a truth which finally awakens him from his unconscious state: In life, they tell him, "you either build or destroy."

As I listened to "Exhibit C" over and over again, I began to really appreciate the wisdom in the "jewel" the Five Percenters dropped on Jay. Though I had never thought of it in those terms before, it is undoubtedly true that through our actions and our mentality, in life we are either "builders" or "destroyers." The "builders" are focused on trying to elevate not only themselves, but the world around them (knowing that by elevating others, they elevate themselves). The "destroyers" never awaken from their unconscious behavior and as a result destroy themselves and everything they come into contact with."[*].

---

* Simmons, Russell, *"Super Rich: A Guide To Having it All,"* Gotham Books, NY 2011

"One for All," Brand Nubian's first and controversial record was released on December 4, 1990. The group was not only inspired by the teachings of the Gods and Earths, the members were and continue to be members of the Nation of Gods and Earths. On July 7th, 2014, I met with actor and recording artist, Lord Jamar of Brand Nubian to discuss the NGE influence, the turning point of Hip Hop, and how this unique genre of music could possibly return to its conscious roots:

Starmel Allah (SA): Peace God!

Lord Jamar (LJ): Peace to the God!

SA: A recent event that went down was the Brooklyn Hip Hop festival. What was that experience like, God?

LJ: I thought it was Peace! I was looking forward to seeing the brother Jay Electronica. When he showed up with the Fruit of Islam (F.O.I.) like that. Well actually when I got there I saw the F.O.I. were there and they were saying that they were there for Jay Electronica, I said okay. When he came, he came with a swarm of F.O.I. I didn't know what time he was coming, I said okay, I thought that was Peace. We haven't seen (especially in Hip Hop) any kind of positive example as a black man being represented in quite some time.

SA: It was something different?

LJ: Yeah, definitely was something different. I feel that kind of positivity can be used right now in Hip Hop and you know, a lot of the things that were dominant and help shape and mold Hip Hop. Especially the Gods and Earths, we haven't really been in the lime light like we use to be as of lately. So with that being said, again I felt it was positive to see Jay Electronica representing the NOI, you know he had his suit on and all that. They were all moving as one and looking sharp and strong. Again, I felt like that was Peace. We ended performing (Brand Nubians), it was great; response was great, the whole crowd response was great. Later on when he performed, I didn't witness this, but at some point, he had the chain Jay Z has with the Universal Flag on it, the one he purchased with his own money, that is his property (whether we like it or not). (*chuckles)

SA: (chuckles)

LJ: Later on he (Jay Z) pretty much bestowed the chain to Jay Electronica on the stage, but when I saw him again (because I saw him earlier) he greeted me, and then I saw him again they wanted us to take pictures. I ended up taking a picture with him and I saw he had the flag on. My thought was oh okay, Jay let him wear the chain or something for a little while today. I really didn't know what the science was behind it with all the commotion going. That wasn't the time or the place to start trying to grill somebody about the purpose of the chain, or why they have the chain. I just did the knowledge to it.

SA: Right! That is what the civilized is supposed to do.

LJ: Yea, that's what you do first; you look, listen, observe, also respect. If I was to do wisdom knowledge that brings about a bad understanding. I did the knowledge to what was going on. At no time did I come in contact with Jay Z. That day, I would have had to really try to follow him and chase him down and I don't do that for nobody.

SA: You aren't supposed to, God.

LJ: Yeah, so you know he was surrounded by a lot of people. The only reason I got to him is because he spotted me and he came to me, or else I wouldn't have been able to get to him either by the way he was surrounded. So, I say all that to say, I'm just telling my story and experience of what I went through that day. When people saw the picture of us, there were a lot of people who had a problem with it. All I can state to that is everybody has their own understanding of how they see things. You can't expect me to react the way you would react. You see what I'm saying.

SA: Exactly

LJ: If you act a certain way in your cipher, then that's how you do it. But I didn't look at it and have any sort of problem with it at the time. So never in my mind would I think to snatch this off his neck.

SA: That wouldn't have been right and exact anyway.

LJ: Of course that wouldn't have been right and exact. We are not here to be robbing our brothers and sisters, you know what I mean. I feel like Jay Electronica (and I'm talking about Jay Electronica), I'm not talking about Wakeel. A lot of people act like they got problems with the brother Wakeel. I don't know Wakeel like that either, I never really spoken to Wakeel, other than Peace or wassup. I have also read sections of his book. That being said, I feel the brother Jay Electronica has been shaped and molded by the different things. He has been shaped and molded by the Gods & Earths, the NOI, the streets, Christianity and Buddhism. I think he wears a lot of these different things on his sleeve and I feel that he wants to represent the different things that has shaped and molded him. Now sometimes these ciphers are going to intertwine with each other. We can let this be teaching opportunities for us. What we need to do is do our duties and stop worrying about what someone else is doing.

SA: That's right. The builders are going to build. That's what it's for, the opportunity for brother to build and teach the people out there that's wondering what that flag is about. There's your opportunity right there to raise the dead or to as we say give the light to those who are in the darkness. Just bringing it back a little bit to what you mentioned earlier about Hip Hop. It being a positive influence and a positive gesture, something that Hip Hop hasn't really seen in quite a while. This is concerning the state of Hip Hop — I remember watching a video on VLAD TV, you spoke about Hip Hop being our house. We are the originators of it and other folks (white in particular) being a guest in our house; and I get that message real clear. But when it comes to the nation would this apply to the nation as well? We are the originator of it and what we are seeing is more and more people (because of Hip Hop, white people in particular) who are interested in becoming a part of the nation per se. We could end up seeing a similar trend where people start participating in the culture and next thing you know it's almost like a takeover type of issue. How do you see that?

LJ: You know, this is why we make them study from 35-50 years before they can learn and do like the original man. We have certain laws prescribed to make sure that somebody is really about what they say they're about; and that's where that science comes from. The problem with Hip Hop is that we didn't make these white/Caucasians that come in (to try to rhyme); we didn't make them study from 35-50 years (SYMBOLICALLY). That's also understanding power to power cipher. People like Eminem, I consider him a type to study from 35-50 years.

SA: He did his homework as far as Hip Hop goes.

LJ: You have to learn to do like the original man. But now you got white boys out here that don't care about the culture. They didn't study from 35-50 years, because within that 35-50 years, you learn who the original man is, you learn who you are. So you know your place in this universe and you know that you are NOT the Supreme Being black man, you know you're NOT supreme in this Hip Hop shit. You're just a carbon copy; you're imitated just like on the planet earth. So Hip Hop is a microcosm of what's going on in the world and within our nation we have prescribed laws, but we have to make sure that we enforce these laws.

SA: Right and Exact

LJ: A lot of *Internet Gods* are out here now. Shit, I can join a group on one of these websites and they'll send me some lessons for joining.

SA: You were telling me the day we were in the studio about that incident. The person didn't even know who you were.

LJ: They didn't know who I was at all. I joined some group on yahoo and they sent me some lessons. It wasn't a full set of 120, but it was a nice amount in there. It definitely was shady. That's not how we do things; that's not how we administer our lessons. So, that being said we have to be better teachers and make sure that we use our teaching methods properly. Yes, we teach knowledge, wisdom & understanding to all human families on the planet earth; we can even teach mankind as well. Again, we have to study from 35-50 years.

SA: That's peace. I just wanted to go back a little to Hip Hop per se. Can you tell us little bit about your first experience with Hip Hop? And who were some of the brothers that influenced you? I know that you have been through this question quite a few times already. But for the purposes of this book God, I just wanted to go into the God's influence on Hip Hop.

LJ: Well, for me I guess the early Gods would be Raheem from the Furious Five; he was the first God, but I didn't know he was the God at that time. At the same time he had that magnetic. You know Raheem, I just like his name, his name was attractive, and he had a smooth style. The ones that really attracted me to this was the Supreme Team on a Hip Hop level. It wasn't til later that I found out that they were teaching and would teach you. They're the ones that sparked the fire in me and planted the seed in me in order to do the same thing. I felt like what they did was genius. I just loved them on a Hip Hop level as DJ's, MC's and the creativity they had when they were making their own original songs. So when they sang 'The Enlightener', I didn't know what that was. I just thought it was a song they made up and I thought it was catchy. I knew all the words; I had no idea what they were talking about. Even when they were on their talking about "Peace to the Gods," I didn't really know what they were talking about. I was very young at the time. So later when I finally met somebody who was God, my enlightener True King Allah from the Desert; when he started giving me lessons, I'm like wait a minute, I know this. This is very familiar. I had already read Message to the Blackman and stuff, but when I started seeing the math and all that, that's when the light bulb went on. I said "Ohh, that's what they were talking about!" I was like "Wow, that's crazy!" They actually made it easier for me to memorize my math, you know, because I already knew it. Without even knowing it I knew it. I was already into Hip Hop and then once you get knowledge you get that fire when you're a newborn and you just want to teach everybody. So I was like "Man, I need to do what the Supreme Team did; I need to put this science in the rhymes in some kind of way." Back at that time it was kind sticky whether that was ok to do it or not. People kind of felt a way about exposing this knowledge. But, I use to say little things back in the day, but I didn't really go hard like putting the lessons' *lessons* into it until later when we got the record deal. Then, it felt like it was time.

A lot of people had said little things but I said we can go deeper. I love Rakim, I love PRT (Poor Righteous Teachers), I love King Sun, I love Just-Ice, but I was like "Yo, we can go deeper", you know what I mean.

SA: Right.

LJ: We was touching the surface very lightly and that's when we made the decision to incorporate these lessons in a way. Not to be exactly the lessons, but we are going to incorporate more things from the lessons that are familiar to the Gods.

SA: And it worked! Just imagine your verse on 'One for All.'

LJ: Right, exactly. I felt like we were saying things that the Gods can acknowledge and the 85ers can be attracted to but we'll lay it out for you, something you can be attracted to, you know?

SA: Word! That was definitely peace. Where did you see the pivotal aspect in Hip Hop when the Gods brought it to Hip Hop? If you were to put a year on it, what year would that be? When the Gods made that tremendous leap and took it to another level?

LJ: That's probably around '86 (Build Equality). You know, "no tricks in '86, it's time to build." Once Rakim came that's when the Gods were strong. Then Kane came after that and he was with the Gods too, but in a different street king of way. Rakim was street, but like Kane was just that...Brooklyn (chuckles)

SA: Yea, he was raw (chuckles)

LJ: Exactly, he was raw. Yea, that's it around there '86-'87. That was a very strong time for us. And it started from there and lasted a good ten years, up until the mid-90's with Wu Tang. After Wu-Tang is when it started to fade out a little something.

SA: Right. Now at that point what do you think happened, God?

LJ: Of course we know there was a concentrated effort to change the messages in Hip Hop when it came to something positive. That can be charged with the introduction and popularity of NWA. Once that gangster rap and all that stuff started coming into play, those gangster movies that they put out — Boys in the Hood, Menace to Society, all those types of things right there. They just want to make money at first (the record companies) and they didn't care what you were saying at that time. That's why people like us and Public Enemy, Poor Righteous Teachers, all those types of people, even Paris over in the west coast. Those people were able to say certain things; we were able to say certain things because the devils just were greedy, they didn't care about the content nor do I think they really understood what we were saying to the fullest. They just wanted to be in the game. They just wanted to get money. So once they started getting money and comfortable in that space, they started to see. Now they starting to here though, what we are saying about them. Then afterwards it's "Aw fuck it, I'm just going to brush that off." But once they saw that they can make money with that exact opposite message and even make MORE money without having to hear what they feel is negative because it's towards them; they started to just promote that. Just to fund and support these things and blow these things up; use the machine to make those things work and draw the machine back when it's time for the positive stuff to come out. They then instill self-doubt on those positive artists, "Oh what's wrong with me? How come my stuff is not selling?" Because they didn't want it to sell, even when it was selling. When our album came out everybody had that album. They still tried to tell us our album didn't go gold. But when you are controlling the computer program that tells you what the number are, of course I can put something in there for artist and stop it at a certain number. Do you think we are really stupid?!? They do think we are stupid, that' the thing. Once they realized they could make money with a negative (what's negative to us) with a detrimental message, that phased us out.

SA: We see the decline in substance in lyrics today. What are your views on that? What can we as listeners, as fans do to create some king of revolution to say "Hey, we want the content to come back. The content of the 80's when the Gods were there." What can we do as fans and listeners to help bring that back?

LJ: Well, the answer to both questions is the same. It goes to EDUCATION! You can't bring content it you have no content. You see, if you're not studying anything or reading anything, how are you going to tell anybody anything. As the consumer, if you're not studying or reading anything, how are you going to demand content that is commensurate of your knowledge of your or the level that you're at. If we're all at a lower level consciously and we just accept strip club music, drug music and shoot 'em up music, then that's what we're going to have If the people educate themselves, they're not gonna want to hear that type of music. They're going to demand to hear something that coincides with what they've been studying. So it's in the devils best interest to keep everybody a part from their social equality; everybody with the lack of the knowledge of themselves in order to perpetuate this, to keep this going. So education on both sides, the artist and the consumer is key. I'm not saying there's one type of education. I don't need everybody necessarily to be within the 5% Nation, Nation of Gods and Earths. Whatever education is good for you that helps the advancement of our people that have been oppressed; has had the knowledge stolen from us; melanated people around this planet earth. That's what I'm riding with. I love my nation, but I don't expect us all to fall within this nation. There are many different ciphers (an infinite amount of ciphers) and we have to not be scared to come together with some of these ciphers and work together in harmony in order to come together with what we have in common.

SA: That's peace, God. I think you pretty much just summoned it all up in that answer, God

LJ: I appreciate it!

SA: Word is Bond! That's it right there. So with that Almighty, I want to thank you. It was definitely real. Peace God!

LJ: Absolutely. Peace God!

DJ WISE DROPPIN' MORE KNOWLEDGE

On July 11, 2014, we reached Washington, D.C. to speak about the NGE influence on Hip Hop at Howard University's Blackburn Center the next day. I was on a panel alongside Akeem Rashad Allah ("DJ Wise"), Queen Divinity, Rashad Sun, Salim Adolfo, and DJ OneLuv Muhammad. The National Black United Front was hosting a National Youth Summit on the same day at the Blackburn Center and we would be there in the mix of it all. When we arrived, we eventually met with Gods and Earths from the region. What we gave was our experiences and knowledge about Hip Hop dating back to its earliest beginnings. Chop Squad DJ's DJ Wise brought records from the 80s for demonstration and most of what he shared there, I thought it would be a good idea to share here.

Before the Gods blessed the microphone, the first element mastered in Hip Hop was deejaying. The common narrative is Hip Hop started in the South Bronx between 1973-74, however as illustrated in the award-winning documentary *"Founding Fathers"* hosted by **Public Enemy**'s **Chuck D**, Hip Hop historians and pioneers such as **Fab 5 Freddie** (Graffiti legend/Yo! MTV Raps) reveal that Hip Hop's origins can be traced to Harlem, Queens and Brooklyn before the Bronx. And amongst these "Founding Fathers" pioneering DJ crews there were Gods in the mix such as **Nu-Sounds** (**Peace Allah**), **Infinity Machine** (**DJ Divine**, **Young God**) and the original **Cipher Sounds**. As Hip

Hop culture evolved and expanded into the mass media, one of the first Hip Hop radio shows in the New York tri-state area also brought Supreme Mathematics to the airwaves and changed the game.

The World Famous Supreme Team Show began broadcasting in 1979 on WHBI 105.9FM, based in Newark, New Jersey. **Sedivine the Mastermind** and **Just Allah the Superstar** (currently known as *JazzyJust the Superstar*) introduced the emerging Hip Hop audience not only to the latest innovative Hip Hop singles but also infused Hip Hop slang along with NGE terminology and Five Percenter ideology. The World Famous Supreme Team's theme song was the Nation of Gods and Earths' national anthem *"The Enlightener."* Their groundbreaking show helped magnify the Nation of Gods and Earth's reach, influence and popularity. The World Famous Supreme Team's success attracted British music producers Malcolm McLaren and The Art of Noise to collaborate and create classic Hip Hop singles such as *"Buffalo Gals"* and *"Hey DJ."* Before **DJ Kay Slay** was known to the Hip Hop community as a mixtape legend, he was a graffiti legend known as *Dezzy Dez* as well as graffiti legend **Kase2** who were Five Percenters as featured in the 1983 classic graffiti documentary *"Style Wars."*

The Nation of Gods and Earths' impact on Hip Hop in every aspect is immeasurable because so many members of our great Nation and those individual influenced by NGE have interwoven our divine culture within the tapestry of Hip Hop's elements.

Much of what we commonly refer to as Hip Hop terminology, expressions, fashion and overall style originates from NGE and was adopted by Hip Hop participants, and through the dissemination of Hip Hop culture, it became popularized to the masses. *"The Fresh Prince"* **Will Smith** helped to popularize the phrase *"Peace Out"* to his *mainstream "Fresh Prince of Bel Air"* television audience, however this derives from the universal salutation of "**Peace**" in which members exchange positive energy upon greeting and departing each another. The iconic *"B-Boy Stance"* actually derives from Gods and Earths "**Standing on the Square**", in which the right arm is over the left forming a square; reflecting one being balanced on every angle, standing perpendicular to the square at ninety degrees (symbolizing being upright, civilized, honorable and Godly), and your cipher being complete at three hundred-sixty degrees (90 x 4 = 360). Because so many Gods were also B-Boys ("Boogie Boys"/ Break Dancers), Deejays and Emcees, it was adopted and referred by many merely as a Hip Hop mannerism. Other examples NGE terminology infused in Hip Hop are: *"word", "word is bond", "maintaining", "dropping science", "dropping jewels", "dropping bombs"*, etc.

The powerful magnetism of the Five Percenter's teachings and its strong influence on Hip Hop culture was highlighted by *"the Bible of Hip Hop"* **The Source Magazine**'s March/April 1991 edition (issue #19). The cover story *"An Islamic Summit: Righteous Rappers Talk About Hip Hop*

and Islam" written by renown Hip Hop journalist **Harry Allen** covered a unity roundtable discussion amongst popular emcees **Big Daddy Kane** and **Lakim Shabazz** representing Nation of Gods and Earths, and **Paris** representing Nation of Islam. If one asks a Hip Hop artist or Hip Hop historian who are the top 20 greatest emcees of all-time without question there will be several emcees listed who are either current members of NGE, former members of NGE or influenced by NGE due to impact of Nation of Gods and Earths during the *"True School"* era (1973 – 1985) and *"Golden Age"* era (1986 – 1995) including but not limited to: **Big Daddy Kane**, **Brand Nubian**, **Poor Righteous Teachers**, **Lakim Shabazz**, **Queen Latifah**, **X-Clan**, **KRS-One**, **LL Cool J** (formerly known as *Lord Supreme Shalik*), **NaS**, **AZ**, **Busta Rhymes** (formerly known as *Tahiem Allah*), **Guru** (**Gang Starr**), **Black Thought** (**The Roots**), **Large Professor**, **Dead Prez**, **Ras Kass**, **A Tribe Called Quest**, etc. Typically topping off everyone's list of Greatest Emcee is *"The God Emcee"* – **Rakim Allah**. Even in the world of Pop-Dance music NGE's presence was felt with 1990's multi-platinum group **C+C Music Factory** featuring rapper **Freedom Williams**. Due to our in depth studies, one's vocabulary and thoughts expand so it's no coincidence that NGE emcees are responsible for revolutionizing Hip Hop's lyrical structure, complexity, delivery, concepts and overall style.

RZA (**Prince Rakeem/Ruler Zig Zag Zig Allah**) exemplifies the application of NGE tenets within the music industry.

After a brief stint as a solo artist, *"The Abbott"* RZA formed **Wu-Tang Clan**, a nine-man emcee collective comprised of: RZA, **GZA/The Genius (Justice Allah)**, **Ol' Dirty Bastard (Ason Unique Allah)**, **Method Man (Shaquan God Allah)**, **Ghostface Killah**, **Raekwon the Chef**, **Inspectah Deck**, **Masta Killa (Jamel Irief Allah)** and **U-God**, along with **DJ Allah Mathematics**. Family members RZA, GZA, O.D.B. and **"Popa Wu" Freedom Allah** introduced NGE teachings to the group members as they came together for one common cause — to overtake the record industry by establishing Wu-Tang as a super group with their debut album *"Enter The Wu-Tang: 36 Chambers"* and then spin off into as many side projects as possible. RZA revolutionized Hip Hop not only through innovative music but also with business practices.

Through our 120 Lesson we learn to become *"The Maker and The Owner."* Wu-Tang Clan was one of the first Hip Hop entities to take destiny in their hands by creating their own independent record label and management company when they released the Hip Hop classic single *"Protect Ya Neck."* The single's success bestowed RZA the ability to leverage a recording contract allowing the group to own their recording masters and enabling members the ability to freely sign independent solo recordings and collaborate with other artists — which was not common practice at that time. Wu-Tang Clan were one of the first Hip Hop acts to truly understand the importance of cultivating their brand and capitalizing on their worldwide appeal with synergetic businesses such as *Wu-Wear* clothing line, *Wu-Wear* retail store, television and film acting roles, music video/film directing, producing major film releases and orchestrating film scores. These ventures allowed Wu-Tang to employ and mentor others.

RZA's application of Mathematics transformed the lives of his group members from poverty and crime into becoming successful artists, established businessmen within the entertainment industry, spiritually and financially enriching their respective families as well as their communities. Through RZA's vision, he showed and proved that a determined idea, planning, execution, and a righteous way of life are the key ingredients towards achieving the **12 Jewels of Islam**. – *Akeem Rashad Allah ("DJ Wise")*

L TO R: Born King Allah, DJ Wise (Akeem Rashad Allah), Hakim Green (Channel Live), Lord Jamar Allah (Brand Nubian), Kasim Allah, Rashad Sun, Justice Allah, and Ness (A-Alikes)

CHAPTER 8

Back to the Basics

As with any group of conscious people, whatever conscious ideology we may subscribe to, we must act on building institutions around that ideology. We may have expert teachers at teaching how the Sun holds these planets in orbit, but every nation needs infrastructure, an economic system and elected officials to represent that nation in political and economic affairs. Unity is important in making the systems we set up work. If you want to see a nation decline in the blink of an eye, do not set up a national economy, do not practice group economics, do not support its producers, and most importantly, spend your money outside your community. Some spend an exorbitant amount of time on history, but not in the sciences and economics. Effective networking, organizing, building structure, voting on processes and procedures are important and shouldn't be neglected.

Elamjad Born Allah, one of the nation's earliest journalists who co-founded The Word and The Five Percenter newspapers, wrote a compelling article entitled "Unification is the Directive, Truth is the Criteria," in The NGE Power, a newspaper that he also founded and was the Editor-In-Chief. The article focused on matters of national import. Elamjad raised significant concerns that anyone desiring to seriously help build a nation should consider. The message is just as important now as it was when it was written back then. Elamjad stated:

Our *"nation"*. In a material sense, what is that? Who is that? Meaning who are its members? What is the census? What are the criteria for identifying them to the extent of good standing? That is their mandatory obligation to the collective. What is their economic duty, their military obligation to service? What is their function in a decision-making capacity and or hierarchy? What are the shared resources of the nation, save Allah School, if that? Who says who is, and who isn't? Also, what is the basis and means of enforcement? Ultimately who specifically, and in what numbers, shall be obligated to, and inure to the benefits of such efforts once accomplished? Moreover how are these processes achieved, over what time period? Is there a progressive evolution perhaps, or some other installation or infusion of the requisite systems? Absent of the tangible usability of answers to such queries, we shall be loath to proceed successfully. Furthermore, this reality applies to any number of projects whose scope and magnitude encompasses the entire "nation." This is serious business, perhaps the most serious in uncharted regions. It is only my intention to elucidate, even in some small sense, some of the preliminary work required and challenges to surmount as one would enjoin this magnanimous nation-building endeavor. It is not my purpose to cast doubt, but rather to indicate the importance of thorough assessment and the consequent preparations. These are absolute imperatives if the matter before us is to be given any serious attention.*

The questions posed by Elamjad present a universal concern for many who have become a part of this culture over the years. A common short answer given to many of these questions depends on who you ask. Sometimes, who you ask isn't the person actually doing the work. The reality of having questions and answers without action presents only more unanswered questions. Those who qualified themselves to act on such tasks led the way but were either met with local resistance or short-lived support. The time is upon those who are the best qualified volunteers to pick up where others left off, learn from other's mistakes, and continue building with a renewed determination for success on personal and national levels. Before anyone can properly build, they have to first know how to read a basic blueprint.

"Life and death are in the power of the tongue." We studied in the lessons that our word is bond and bond is life, and we will give our life before our word fails. Now, that's a dynamic principle that shapes the moral character of people *to be doers* of their word. The character of a person who will give their life before their word fails is a person that will do whatever is necessary or required to make their word true. This means there are no excuses especially if you know better. Much has been said over a fifty year period, but how many did what they said they were going to do? I experience this often. Someone says they will call and they don't. Someone says they will follow through on a project

---

* See, Appendix for full article: **Elamjad Born Allah, *The NGE Power*, Issue #1.4, April 2003, p. 2**

and they don't. Someone says they will get something important finished by a particular time and they don't. How do you feel about someone who says they will do something and they don't do it?

If someone says they're going to study their lessons, but they never do or always seem to come up with an excuse for not studying, that person isn't even qualifying themselves to know our most basic teachings. If they can't commit to knowing the basics, how can they handle what's to come in the future? This is why some of us need to get back to the basics now. Some of us really need to take it back to the foundation because somewhere along the way, we seemed to have forgotten or missed some important things about what we are to be building. The basics — Supreme Mathematics, Supreme Alphabets, 120 Lessons, general discipline, etiquette, and honor — are integral to who we are, what we say, and what we do. Our success depends on our ability to follow through on what we say we are going to do. I have been in many circles exchanging ideas and sharing opinions back and forth. I soon learned without a plan and people committed to doing what they said they are going to do, the ideas and opinions would be in complete vain. The more people that can follow through on a planned course of action, the more successful they will eventually become. Just as our lessons referred to applicable scripture, the following passage is the most fitting not as a religious reference but because of its scientific correlation:

> *"As the rain and the snow come down from heaven, and do not return to it without watering the earth and making it bud and flourish, so that it yields seed for the sower and bread for the eater, so is my word that goes out from my mouth: It will not return to me empty, but will accomplish what I desire and achieve the purpose for which I sent it."*
>
> —Isaiah 55:10-11

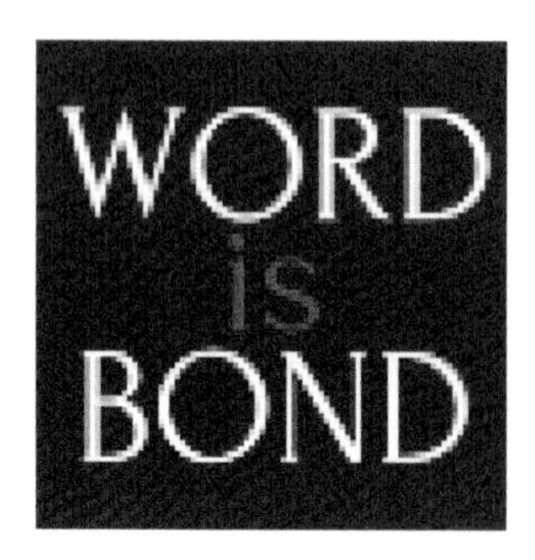

FIVE HONORABLE REASONS TO LIVE UP TO YOUR WORD

1. Integrity

For some people, talk is cheap. For us, Word is Bond. For me, making your word bond is a matter of *integrity*. In the beginning was the Word and the Word was with God. The Word was/is God and that Word became flesh which is a material reality. I wouldn't feel good about myself if I didn't keep my word. My word means everything to me — I do not take it lightly. I don't make promises I may not be able to keep. There is a lesson we studied and recited which states for peace and happiness, "I will give all I have and all within my power." Some of us took that personally and that became the standard we applied to our word. If we said we were going to do something, we gave all we had and all within our power to see our word become a reality regardless to whom or what. If your keeping your word doesn't mean much to you, why is that? And more importantly, what does it say about you?

2. Trust and Reliability

No bank or government would ever work if they didn't do what they said they would do — protect your money. We put our trust in banks to save our money. We put our trust in the government to maintain the well-being of the nation. We are not naïve to corruption, but for the most part, we trust the law to bring justice to the wrong doers. Without any kind of trust, nothing will ever work. Usually, our trust is given based on the perceived reliability of someone or something. I do not trust people who don't keep their word. If someone lets me down a number of times, then I know they are not reliable. I find that this limits my desire to spend money or time with them, which is sometimes sad, but I have learned to accept that I cannot trust them to follow through on what they say they are going to do. The success of anything thrives on trust and reliability.

3. Respect

If your word doesn't mean much to you, people will not respect you. It's a letdown when someone says they are going to do something and then doesn't do it. Of course, I'm the type of person to give another leeway but not that much. There may be a good reason they didn't do what they said they were going to do. But if it happens over and over, I accept that I can't rely on them and may not take their word seriously. If you want to feel respected by others, then you need to say yes when you mean yes and no when you mean no, and not allow your fear of rejection or your fear of being controlled to get in the way of being a trustworthy person. Refer back to point number 2 about Trust and Reliability.

4. Self-Worth

We cannot feel worthy when we let ourselves down by letting others down. People who renege on their word do not value themselves enough to act with integrity. A positive self-image and self-esteem are key to character and integrity. The value placed on keeping our word should be a central aspect of our self-worth because it is who and what we are. Whether we have high values and principles or low ones, they tend to tell people who we are and how to treat us. Do you feel like a worthy person when you don't keep your commitments? Do you value people that don't keep their word? Self-worth is the result of treating ourselves and others with care and respect.

5. Personal Power

Personal power is the result of behaving in ways we value. People of great personal power would rather die than compromise their integrity. People who will give (not take) their life before their word fails tend to exhibit a great ability to make things happen. They will scrape and scrounge, sacrifice whatever is necessary to achieve something great. Potential power turns into a kinetic response into action when people do (not say) what is necessary. Some people would rather die before compromising their integrity, their word, or anything dear to them. All people have this kind of power in them but we use it for different things. Some of us tap into this kind of power for the right reasons and others for the wrong reasons. Fear and doubt blinds us to this power.

THE BLAME GAME

By raising standards for yourself, you begin to make the change you are looking for. Change happens within you first. Where does it happen? In your mind: Your thoughts and attitudes. How does it happen? Replacing old thoughts and attitudes with knowledge and let that knowledge serve you. When does it happen? Now! You don't have to take no for an answer and you don't have to keep

accepting the norm or waiting for someone else to bring about the change you want for yourself. Be the change you are looking for by being the example of what's missing. Take charge and take the lead.

In this culture, some were given the same lessons and some were not. Some were taught differently and some were taught the same. Some were taught by the best and were given the best of wisdom while others were given the worst teachings by the worst so-called teachers. America is the most culturally diverse place on the planet. People bring their cultures with them to America and some relearn their cultures here. Through acculturation, some adopt or abandon other cultures. We are all here mixed in together. When it comes to this culture, many came and went. There are those who chose to remain a part of this culture while others chose to go a different way. Some joined a church while others joined a temple or mosque. Allah's position on this was quite liberal as he knew some would need the discipline the Temple or Mosque offered.

It should come as no surprise when someone chooses to part ways with this culture. People left the Mosque and joined the NGE and people have left the NGE and joined the Mosque. Where people go is their choice not mine, therefore I don't waste time debating the issue. Dihoo Supreme, is the reverend of Chambers Baptist Church in Harlem. He's held funeral services for Five Percenters at his church which is a staple in the Harlem community. After learning the lessons, he told Allah he had to go his own way. We've been teaching for years that *"Many shall come but few are chosen. Those who are chosen shall choose themselves."* People have a choice whether to stay and help build this house or leave. It's all a matter of choice. We all choose what's best for us at the time.

For years, months and days, we have attended Rally after Rally, Parliament after Parliament. We have seen beautiful ideas become born and we've seen ideas get aborted. Some things have changed but some things stayed the same. Some of us have gone astray and some of us have returned home. We have seen babies saved and babies killed. We have seen elders return to the essence of life and newborns just starting life. We have learned and continue to teach the science of everything in life but there is so much frustration and stress that comes with how we build in this life. Why do you think that is? Why do people walk away and fall back from this thing? Why do people get frustrated over this thing, go into self-exile and become its greatest critics? What makes people stay to their selves and not want to deal with social, economic or political equality? What brought about their dissatisfaction? Is it the complacency or stubbornness of others? Or could it be that the opinions and approval of others matter so much to some, that the wrong opinion or lack of support causes disappointment, discouragement and ultimately, resentment?

The previously mentioned examples are various reasons why brotherhood and sisterhood aren't the same as they used to be, why morale appears low, and why people are less enthused about unity and building, today. People rather sit back, criticize and judge others instead of lending a helping hand, volunteering, or making some small contribution (whether financial or otherwise) to a good cause. In the land of the know-it-alls, why do pessimistic views run so rampant and why do we let it kill our enthusiasm? What happened to all those Five Percenter songs? You know something is up when people no longer sing or teach their children to sing their own music (and I am not referring to "The Enlightener"!). Did we get too old for that? Did we outgrow certain aspects of our culture? Just imagine how much an entire generation will never know because people stopped teaching. Please do not confuse teaching with regurgitation. Teaching is always fresh and relevant to the time whereas regurgitation becomes blasé, stale, corny, old and out of touch with changing times. We teach and show and prove by example but there has been much regurgitation along the way. Could it be this is why Allah didn't want to get old?

CHAPTER 9

The Right Direction

What is movement? Many writers have described Allah's Five Percent Nation of Gods & Earths as a "movement" or "youth-based movement." A movement is just a group of people working together to advance their shared political, social, or artistic ideas.

Everything in the universe moves and has motion. Even if it looks like it is not moving it is moving. This is because one of the laws of nature is motion. Now, look, I have to get a little scientific with ya'll, stay with me. In physics, motion is a change in position of an object with respect to time and its reference point. Motion is typically described in terms of displacement, direction, velocity, acceleration, and time. Motion is observed by attaching a frame of reference to a body and measuring its change in position relative to that frame.

If the position of a body is not changing with the time with respect to a given frame of reference the body is said to be at rest, motionless, immobile, stationary, or to have constant (time-invariant) position. An object's motion cannot change unless it is acted upon by a force as described by Newton's first law. Momentum is a quantity which is used for measuring motion of an object. An object's momentum is directly related to the object's mass and velocity, and the total momentum of all objects in a closed system (one not affected by external forces) does not change with time, as described by the law of conservation of momentum. The study of motion deals with (1) The study of motion of solids (mechanics) and (2) The study of motion of fluids (fluid mechanics).

As there is no absolute frame of reference, absolute motion cannot be determined. Thus, everything in the universe can be considered to be moving.* Spacetime (the fabric of the universe) is actually expanding. According to the latest science information, everything in the universe is stretching like a rubber band. This motion is the most obscure as it is not physical motion as such, but rather a change in the very nature of the universe. The primary source of verification of this expansion was provided by Edwin Hubble who demonstrated that all galaxies and distant astronomical objects were moving away from us ("Hubble's law") as predicted by a universal expansion.

The Milky Way galaxy is hurtling through space at an incredible speed. Many astronomers believe the Milky Way is moving at approximately 600 km/s relative to the observed locations of other nearby galaxies. The Milky Way is rotating

---

* Tyson, Neil de Grasse; Charles Tsun-Chu Liu; Robert Irion (2000). *The Universe: At Home in the Cosmos.* Washington, DC: National Academy Press

around its dense galactic center, thus the sun is moving in a circle within the galaxy's gravity. All planets and their moons move with the sun making the sun appear to be still, but the solar system is moving.

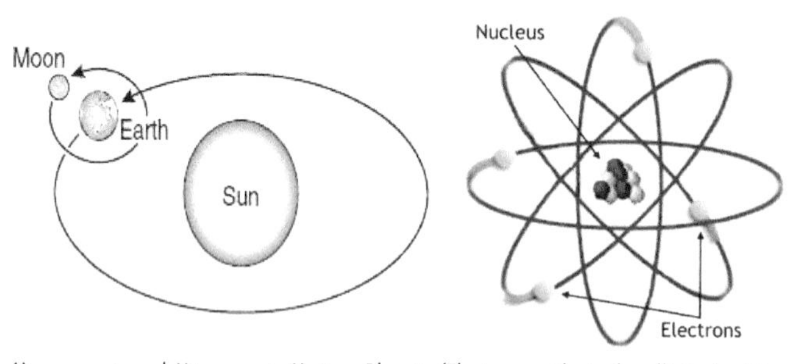

Macrocosmic and Microcosmic Motion: Planets/Electrons with similar elliptical orbits.

The Earth is rotating or spinning on its axis, this is evidenced by what we call day and night. The Earth is orbiting around the Sun in an elliptical revolution. A complete orbit around the sun takes one year or about 365 days. The Theory of Plate Tectonics tells us that the continents are drifting on convection currents within the mantle causing them to move across the surface of the planet.

The human heart is constantly contracting to move blood throughout the body. Through larger veins and arteries in the body our blood travels. The air we breathe and the foods we eat also move about. The smooth muscles of hollow internal organs are moving. Sound is audible at any given moment and when the vibration of these sound waves reaches the ear drum it moves in response and allows the sense of hearing. The human lymphatic system is constantly moving excess fluids, lipids, and immune system related products around the body.

The cells of the human body have many structures which move throughout them. Cytoplasmic streaming is a way which cells move molecular substances throughout the cytoplasm. Various motor proteins work as molecular motors within a cell and move along the surface of various microtubules. Motor proteins are typically powered by the hydrolysis of adenosine triphosphate (ATP), and convert chemical energy into mechanical work.

According to the laws of thermodynamics all particles of matter are in constant random motion as long as the temperature is above absolute zero. Thus the molecules and atoms which make up the human body are vibrating, colliding, and moving. This motion can be detected as temperature; higher temperatures, which represent greater kinetic energy in the particles, feel warm to humans sensing the thermal energy transferring from the object being touched to their nerves. Similarly, when lower temperature objects are touched, the senses perceive the transfer of heat away from the body as feeling cold.

Within each atom, electrons exist in an area around the nucleus. This area is called the electron cloud. Electrons move in strict paths the same way planets orbit the sun. They have a high velocity and the larger the nucleus they are orbiting the faster they would need to move. Light moves at 186,000 miles per second. The evidence points to the fact that everything in the universe has a movement unless it is declared dead or motionless. Human interaction abides by these laws and involves motion also. If people have an individual or collective problem, some kind of movement is needed to solve it. Problems do not work themselves out mysteriously or without some kind of force creating motion to change it. As with all movements, people are required to do work (force) to change a condition or solve a problem. If no one does any work, it is likely the problem will stay the same until some kind of force is exerted.

Why is it necessary to explain this scientifically? It is necessary because everything moves in the right direction (or in the direction it's supposed to move in). Everything in the universe *moves with a purpose.* We must learn to be in-tune with ourselves to the point where we are moving with purpose and not aimlessly. This requires us to be fully aware (have knowledge) of ourselves, and be able to put that knowledge to work *with a purpose.* We need to understand **work** (not just wishing or hoping) is necessary behind any idea or movement to bring about life, progress or solutions. We all got here because two people had to *work it*! Our time and energy should be used to *do something* about our problems instead of wishfully thinking, complaining or criticizing others. So what direction can we move into with a purpose?

SPECIAL S.O.S (Save Our School) EDITION

PULL OUT

THE FIVE PERCENTER

Issue #2 September 1995

"Our 1st Goal Is Land"

LAND & ADVERSE POSSESSION

The Five Percenter Newspaper, Issue #2, September 1995

THE LAND, LORD!

South African leader, Desmond Tutu once said, *"When the missionaries came to Africa they had the Bible and we had the land. They said 'Let us pray.' We closed our eyes. When we opened them we had the Bible and they had the land."* Native Americans faced a similar reality in South and North America. After getting knowledge of self, many of us put down those Bibles, but need to come together and get some land (or property) of our own to build on. With gentrification targeting our neighborhoods, this is an issue that can't be ignored or side-stepped any longer.

The lessons we are all required to study give us actual facts relating to land and its use by the total population of the Earth. In order for a nation to build, that nation needs land or property to build on. It makes sense then to make land or property acquisition a priority because land or property is needed for us to build a world for ourselves. Useful land is only useful when people are actually using land to make a life for themselves. Otherwise, the land will just sit there waiting to be acquired by someone else. The acquisition and transferability of land to the next generation by qualified and reliable people is achievable with collective work and consistency. It is a process that has no room for ideological differences, but has all room for like-minded, dedicated people who are ready, willing and able to fix their credit and pool their resources together. We have people in our nation right now who are in real estate, banking, finance and law. The Gods and Earths of Power Hill (Philadelphia) have proven that coming together yields great rewards as they've managed to secure the deed to what is now *The Universal Street Academy* (see, *Chapter 21 of The Righteous Way, Vol. 1*). It's time we get together and put these determined ideas in motion or find ourselves out of doors. We have the numbers to make it happen and the next generation will thank us for doing this. So how do we do this? ***1. Get focused! 2. Get a plan! 3. Get a team! 4. Get organized! 5. Get busy!*** That's the simplified version. If you want a more detailed breakdown on how to get organized, please refer to *Chapter 17 of The Righteous Way, Vol. 1.*

CHAPTER 10

Like Minds Attract, Not Repel

THE POWERS OF attraction pull people in and repulsion pushes people away. Understanding how these forces work can allow us to see why COMING TOGETHER with some people is sometimes a challenge. Sometimes people focus on differences instead of similarities (i.e., *Willie Lynch syndrome*). The word "*different*" is mentioned in 120 Lessons in reference to our enemy who really exploits our differences to keep us all from really uniting without compromising the way of life we chose. According to the lessons, we outnumber him 11 to 1 and they are the real minority on the planet yet control the majority socially, economically, and politically. The majority keeps falling for the same old tricks that keep us divided. Sometimes people focus on things that are of no consequence. They forget this isn't about liking people it's about teaching each other, building, and never forgetting the common cause that brought us all together in the first place.

This came to mind as I witnessed the differences between some Gods and some NOI/FOI had online and offline. We don't always have to waste enormous amounts of time and energy trying to persuade and convince each other to accept another's views. There's work that has to be done and our communities wherever we are in serious need of seeing unity, strength, and discipline coming from the Black man and woman. All the people in the community that I have met and spoke with tell me of some bad experience that repelled them from this nation or that nation. The greatest common factor was someone saying they were one thing did the complete opposite. Now, you know what opposites do from what the lessons pointed out. We are the alike attract people not the unalike attract people, and so because our people are righteous by nature, when they see a shameful act, they tend to repel from that person and anything that person is associated with. Energy used incorrectly pushed people away when it should have attracted them here. Abu Shahid, the Father's Brown Seed and elder, posted in an online chat that:

> *"It behooves me to bring to your attention the disruption in the FORCE. There are two parts to one but there is only one. There is only one God, and God is Power and Force. Positive and Negative. Natural, Supernatural and Unnatural, in a kaleidoscope of patterns.*
>
> *There has been a never-ending disturbance in the Force. Electricity is our secret power and its basic principle is magnetism. Unalike attract, alike repel. So instead of our beloved Nation finding Unity in the principle (the seen and the unseen), where our strength truly lies. Diversity is our strength. We're all different, unique in our own particular attributes of perfection. The light side and the dark side. The seen and the unseen.*
>
> *I recently pointed out in a metaphor about the square and the circle. Both being composed of 360 degrees. So tell me why they can't see that Understanding Equality in the Cipher brings peace, harmony and current direction?*
>
> *The spiritual principles of electricity which I repeat are our power, which has two different currents, AC and DC. AC is the Alternating Current and DC is the Direct Current. So I suggest that we accentuate the Positive and ELIMINATE the Negative by focusing on producing Positive Energy Always Creates Elevation, just like Proper Education Always Corrects Errors, and that's PEACE."*

If we find this to be correct, our power and force is diminished by energy spent on trivial matters. We shouldn't forget some currents remain suspended because of shorted circuits. Some circuits run in series while others run parallel. Some of the currents carry high amps and others low amps depending on the conductor. There are various conductors of electricity and each one is different in capacity. Why not respect that and get back to the duty we all said we would do?

The average dictionary defines GOD as the omniscient, omnipresent, and omnipotent being. Let's really look at their language now. The word *"omni"* is a Latin word meaning *"all"* or "every." *Scient* (as in "scientist" *means to know or have knowledge*), *Present* (*means right now*), and *Potent* (*means to be powerful*). Put it all together. Everyone must have knowledge right now to manifest their power. The truth is hidden in the language but is used to make you and I believe it is referring to an invisible creator that no one ever saw. Why? Because the trick is in putting a limit on your imagination by having us rely on something invisible to do for us what we can do for ourselves while they get rich off our labor. All the while we remain blind to who we are, our creative ability and power to make a world we desire.

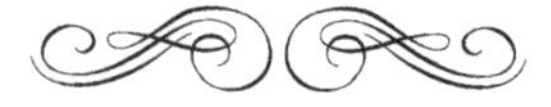

"I can show a religious man that he never led anyone to God. Not the Pope or anyone. They all died and all the people in the Bible died. Now where are they? Where are they? The only way you will find God if you keep on reproducing until he make himself known."

—ALLAH

***OMNI*SCIENT:** The collective knowledge (conscious and subconscious) of past, present, future by the Original Black man, woman and child.

***OMNI*PRESENT:** The total population of original people on the Earth who are everywhere on each continent. The original Mothers and Fathers of civilization and their progeny.

***OMNI*POTENT:** The unity of a people is the great power of a people.

The spoken word of TRUTH can electrify others and therefore has attracting powers in it. Whenever our great orators and leaders spoke, they had the power to attract large groups of people and organize those people. The devil knows the power of attraction which is why he continues daily to teach the 85% that God is a righteous, unseen being that exists everywhere. FBI Director J. Edgar Hoover issued directives under COINTELPRO, ordering FBI agents to "expose, disrupt, misdirect, discredit, or otherwise neutralize" the activities of these movements and their leaders. He was mostly successful as we can see today, but we are still here. That has to count for something.

What are some of us afraid of now? If you have ever seen people come together for one common cause, you know it is the most beautiful and powerful thing. That, my family and friends, is ALLAH (OMNIscient, OMNIpresent, and OMNIpotenet). We must remove all excuses today for coming together and having true unity. When we do not allow excuses to live, we strengthen our magnetic field as it grows each time someone is electrified by the TRUTH and adds on to the cipher.

I found a great lesson about something having attracting power when Allah Forever told me a story one day. He gave his nephew a pin of the Universal Flag and told him to guard it with his life. He told his nephew if he lost it he would bust his ass. Every day, Allah Forever made sure his nephew wore that pin. The friends of his nephew soon took notice of the pin and were curious about it. They desired one also and wanted to find out how to get one. His nephew soon brought all his friends to Allah Forever and he began teaching them. Interestingly, Allah Forever did not have to go searching for students. The students came to him seeking that flag because they were attracted to it.

Our oral tradition in the nation created widespread interest in these teachings and these teachings spread throughout the world because of it. As each one taught one, another heard something powerful enough to attract them to that person who was teaching. The words of the speaker contained enough power to naturally grab the attention of the listener and draw them into the discussion, elevating their minds during that moment above the norms of everyday life. But the word of a God or Earth wasn't the only thing that attracted people. Their ways and actions were characteristically different from the average person. They spoke, ate, and dressed differently and this was a distinguishing trait unique only to the Five Percent.

A person's etiquette can make all the difference in teaching about our culture. As we mature and grow in years and experience we must be able to meet more social situations with professionalism, courtesy, confidence and ease. This should be no problem for anyone who considers themselves all-wise and righteous, a master, or a teacher. How a person speaks and acts towards others will be associated with their name first. Therefore, proper conduct and etiquette should be kept in mind at all times.

Everyone wants and demands respect, but what is it? How do you show it? How do you give it? Most importantly, do we have it for self? Respect means to acknowledge a person, place or thing with value and to treat it in an appropriate way. Some people will say respect has to be earned or built over time. Others will say it is something that is just given by a respectable person. My mother didn't have to earn my respect or build it over time. Respect was something you just did to show good manners or that you had good home training. A respectable person knows how to act at home and abroad.

You don't have to agree with someone or something to show respect. Respect is primarily about the civility of a person and then about how that person shows it to others wherever that person goes. If a person says they are all-wise and civilized, that must be a person that has respect for their selves and others. I have found that the more knowledge a person acquires, the more respectable they are. The gift of knowing also comes with cultivation and a refined state of mind. Isn't the purpose of knowledge of self to also clean yourself up?

Respect is, in my opinion, one of the best exemplifiers of self-control. I have seen discussions go from civilized to uncivilized in less than 60 seconds all because someone lacked respect or forgot that respect is key in communication. I lost self-control plenty of times and with it, I lost respect for where I was, who I was speaking with, and myself. It was not a proud moment to say the least. These days, we are seeing people who have lost respect for their selves, others, and even the teachings that gave them the basis of their knowledge of self. As people with knowledge of self, our etiquette can be seen in just about everything from how we live to how we interact with other people from other cultures, religions, and traditions.

Today, the widespread use of social media causes many discussions to take place online. Many would even admit they were attracted to these teachings from someone they met online. But the online community is open to millions from all walks of life. This openly invites many to see our culture differently from those who were traditionally taught how to live this culture in person. The opinions of those who aren't confirmed Gods or Earths would now somewhat clash with the true and living Gods and Earths who have an online presence. The upside to us having an online presence is that it connects us to the entire world. But the downside is that our teachings have been confused with the teachings of others. Some have even mixed, diluted, tampered, and outright misrepresented our teachings to folks who have no idea who is on the other end of that computer. This consequently causes some people who want to genuinely learn what this culture is about to mistake our teachings for something obscure.

Contrary to popular belief, our teachings are not rooted in the obscure, arcane or mysterious. They are based in the factual simplicities of mathematics, science, nature and history. The goal is to take the mystery out of it and understand its root through proper education. When factual knowledge becomes a person's first law of observation, it will unveil anything shrouded in mystery, leaving only the truth waiting to be discovered and shared.

GO THE WHOLE 9 YARDS: 9 THINGS TO TEACH 9

Today, we can agree that teaching or studying 120 lessons alone does not make a person God or Earth, and teaching one person alone does not fall within the scope of Allah's intent for Five Percenters to go and teach nine others. We have to go the whole nine yards. This is the science of multiplying, repeated additions of a specified number or quantity to increase in numbers. Everyone is a multiplier and the universe is the multiplicand, everything we need is already here, all we have to do is produce. It is the divine right of every human being to be a producer. How do we produce? We have to make something by being the maker. We can achieve this by *using the resources* around us. If there are no resources around you, find where the resources are and use them. That's anywhere on the planet! How can anyone produce something without an education or without having any kind of knowledge? Our teaching is designed to make us productive. This is why Allah

placed education at the top of his priorities and started a Street Academy where Five Percenters can continue their education. A proper education is the right of every human being. To deny a human being an education is to deny a human being of the divine right to live a productive life. Thus, if a human being fails to put into practice what he/she knows, they deny themselves of an opportunity to be productive.

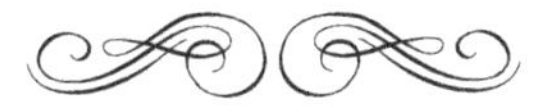

> **"When you go to school you got to have the knowledge before you could come out and put that into practice. And they don't know you got it until you go out and put it into practice. This is why I teach the child to go and get the knowledge, come back and put it into practice, then the people will get the what? Understanding! Then they'll know your culture."**
>
> —ALLAH

One of the earliest Five Percenters to do this was Dumar Wa'de Allah (PBUH). Dumar was directed by Allah to attend Harlem Prep to further his education. Dumar became one of the first Five Percenters to graduate high school and go on to college where he taught fellow students about this culture. Allah wanted each of his students to go out in their travels and teach 9 others which is how the teachings spread so quickly and broadly. Because education is a high priority for Five Percenters, we were made into enthusiastic students who sometimes even challenged the teachers. Some of us were so enthusiastic we got in trouble for disrupting the class.

Today, some mistakenly think the lessons are all that's needed to be who we are and to build. The truth is, however, we need a well-rounded education. The same things we are lacking today are the things we ignored yesterday. A nation needs qualified people in ALL these areas. The lessons are only one part of the curriculum. Students should also have an education in the areas of:

1. **Mathematics** – a pure science dealing with quantities, magnitudes, forms, formulas, distances, measurements, etc. and their relationships to us by use of numbers and letters.

2. **Sciences** – the body of systematized knowledge derived from observation, study and experimentation of phenomena inside ourselves and the universe in which we live.

3. **Language** – The ability to communicate and understand words, their meanings and proper usage in written or oral form.

4. **History** – An account of what may have happened with people, place and things in the past.

5. **Economics** – The management of the income, expenses, etc. of a household, business, community, government or nation.

6. **Agriculture** – The practice of cultivating the land to produce food.

7. **Law** – The rules that maintain peace, order and the rights of people by through justice.

8. **Government** – A structure that balances power through the consent of the people to elect officials among themselves to govern.

9. **Mating** – The proper choosing of male/female partners (Healthy, strong and good breeders).

A basic education lacking in these areas cripples a person's ability to make a productive life for themselves and family. Additionally, it is unwise to think the lessons alone are sufficient to make a productive life. Some of these were not taught to us in school. We were denied or lacked an education in some these areas which is why we have some of the problems we have today. The lessons certainly teach us about these subjects, however even the lessons warn students that there are other rules and regulations *"not mentioned in this lesson."* Back in the day, people got caught up in the wisdom of plus lessons and only wanted to deal with high science, but they didn't want to learn basic things necessary for having a productive life. Too many have gone for themselves without an education of the basics, which is why so many are suffering today. Black Messiah pointed out that a plus degree is getting a G.E.D., a diploma, a trade or a skill because we need qualified and skilled people to build.

CHAPTER 11

Calling to Order

Coming together and making things like this possible for the communities in which we all live is necessary today. One of the simplest things Allah taught was that if we ever wanted to see him all we had to do is come together. If we had any problem or challenge surely Allah is the best knower and is all wise so why not come together to deal with whatever the problem or challenge might be? We come together once a month and many of us come together during our own free time. We started out as mostly teenagers with the knowledge and we all mingled with each other, we engaged each other with the knowledge, we mingled and participated. This is what Allah advocated. We should do the same.

Some are so busy quoting Allah and reminding others what Allah had said they forget to practice what Allah had said theirselves. The vision, courage and initiative of a small few tend to always lead the majority. Within the camp of the quietly comfortable, is the general acceptance of going with the flow as if no organization is needed, no help is required, no responsibility is demanded, and no accountability is necessary. Within fifty years,

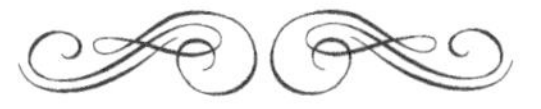

"I said build a nation or be destroyed. And if you stop them [the youth] from mingling together it's over with. It's over."

—ALLAH

our culture has grown from having house rallies to city rallies and parliaments to regional conferences! Now, where do we go from there? What happens next? Many have asked questions about structure and being organized. It's easy to think our answers are *the solution* and we sometimes don't know how to present our ideas to the body or how long it should take.

If it takes the Sun's light to reach the Earth in eight minutes and twenty seconds, surely we can get our point across in that time or less. There's no need to hog up the little time we have

together and putting ideas on the Internet is not the way, sorry. It is necessary to have time limits to permit all speakers to have an opportunity to express their viewpoints, and speakers must restrict their remarks (or announcements) to the items they requested to address. Persons attempting to discuss other matters or to speak out of turn should be ruled out of order and not allowed to speak. I've learned that there's a time to speak and a time for action. Answers and solutions come gradually over time through action, not endless debate.

If we really want change, it's all about bringing people together, sharing information, building relationships and following through on new ideas. This is best achieved when people focus on what makes us similar not what makes us different. Focusing only on ideological or political differences limits people's ability to involve others. On the flipside, others may not know how to get involved.

When a group of people come together for a common cause, that common cause should be clearly stated. It should be understood by others but more importantly, it should be based on *the will of the people.* Sometimes, I see one or two people rallying for support for things that are not the will of the people, it's simply what they want to happen. I remember attending a parliament and a brother couldn't wait for a quorum, he started speaking with only three people present. He had something important to share but was out of order and couldn't wait for others to show up. When others began to show up and fill seats, he said he was leaving. This was the best time for him to be heard, but it turns out he only wanted to be heard by three people, not the nation. What do we do when there is a dispute over the direction the majority should move in? Some might say that's what Parliament is for. What about Parliamentary Procedure?

When people convene for a common cause, there should be a quorum and a fixed agenda. A quorum means the minimum number of members of an assembly or society that must be present at any of its meetings to make the proceedings of that meeting valid. Parliamentary procedure is the set of rules for conducting business at meetings and public gatherings. These procedures are important because they bring order to what otherwise might be an out of control meeting where anyone can rudely talk out of turn, jump up and say whatever they want, or argue over what's important and what isn't. We can set up our own rules to conduct meetings, but if you want a proven method, parliamentary procedure is a smooth method that allows everyone to be heard and to make decisions without confusion. It allows for the whole population (men and women) to participate, not just a few. The meeting is usually called to order, the secretary reads a record of the last meeting, reports from special committees are presented, unfinished business from the last meeting is revisited, new business where new ideas or topics are introduced, announcements and adjournment until the next meeting.

How do others get their say? They make motions or wait until announcements. A motion is a proposal that the assembly can take a stand or take action on an issue. Those in attendance can make motions, second motions, debate (or question) motions, and ultimately vote on motions. The method of voting on a motion depends on the situation and the agreed rules (or bylaws) governing the parliament or meeting. Voting can be executed by ballot (writing your vote on a piece of paper), by a show of hands, by roll call (each member answers "yes" or "no" as their name is called), or general consent (the chair says "If there is no objection..." Members show consent by their silence. If someone says, "I object," the matter must be put to a vote). For efficiency, motions should relate to the business at hand and be presented at the right time (the same goes for announcements). Remember, wisdom out of season bears no fruit! Motions should not be obstructive, frivolous or against general laws or bylaws. Parliamentary procedure helps get things done in an orderly, courteous and respectful fashion. It helps to keep motions in order, helps people to obtain the floor properly, ensures everyone is heard, and helps us to follow the rules of debate when we disagree.

Added to the difficulty of organizing a meeting is the fear that key participants won't attend. We all know that people are busy enough doing their own jobs or attending to family matters, so asking them to join in on something that is beyond the normal can be seen as an imposition. So rather than get turned down, it's easier to not call the meeting in the first place. Don't quit before you start, find out when would be the best time to meet. When things get tough, remember that leadership is a collaborative process, not a position.

Finally, calling a meeting with multiple people (who may or may not be a part of the nation) requires communication and facilitation skills. Getting everyone together is hard enough, but making sure that the meeting is productive, that all parties are heard, and that real solutions emerge is daunting. The truth is that many people have good ideas but lack these skills or may not want to do the work their selves. So for some of you, it may be easier to deal with everyone individually than trying to deal with them as a group. For others, it's easier to give someone an idea to run with rather than put in the work their selves.

Given all of these factors, fears and uncertainties, it's no wonder that many people hesitate to use their convening authority (building in activation). Unfortunately, bringing the right people together from different places (and even including outside organizations and associations) is often the best way to get things done quickly. So if you want to step up to more effective leadership, you may need to conquer your fears and give convening a try. With so many having knowledge on so many levels and egos come out daily like sunshine, we need to know how to: A) *Agree to disagree,* B) *How to agree to disagree without being disagreeable,* C) *How to just agree on basic things so we can get work done.* What do I mean? There's always someone asking "what do you mean," isn't there? I think what I said was pretty clear, but for those who need me to explain a little further, here's what each one means:

A) *Agree to disagree* means to resolve a conflict (usually a debate or quarrel) where all parties tolerate but do not accept the opposing position(s). This usually happens when all sides recognize that further conflict would be unnecessary, ineffective or otherwise a waste of time.

B) *Agree to disagree without being disagreeable* means that neither of you are convinced by the other's opinion, but you respect each other's opinion and agree that you are both entitled to your own thoughts about the subject without falling out with each other.

C) Agreeing means that people share a mutual understanding about a course of action, a situation, or a condition, and about what is to be done.

What we have experienced over the years is that some folks can be real argumentative, disagreeable, or just agree to disagree and things would be left just as that. Some who were able to get things done were the ones who agreed that something had to be done. Our goal now is to start getting people to agree on important matters. This is best done when egos are put to the side. Those of us who know the lessons know about the instances in which "we agreed." The more people can agree on what must be done, the more powerful that group is in making things happen.

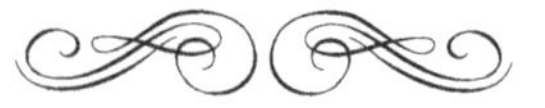

"Unite. Let your children and them unite together and create new ideas. Where the world is building a square now it's going to be building a circle. And they don't need your ideas no more. Your ideas did what you was supposed to do. All you got to give them is the basics because they'll do it them what? Selves! See you can see where ideas have traveled for many years cause when you first come to this country there were no houses. It went from the log cabin to this. So each generation brought this about right? So let the children bring about another new world."

—ALLAH

CHAPTER 12

Doing the Math

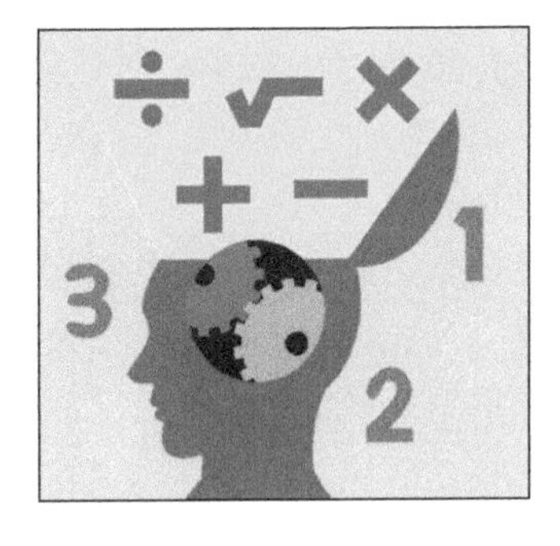

BETWEEN 1930 AND 1934, Master Fard Muhammad taught the Honorable Elijah Muhammad that Islam is Mathematics and Mathematics is Islam and can be proven in no limit of time. He did not say Islam is a religion. He said it is Mathematics. A mathematical mind studies rules, instructions and formulas to solve problems right and exact. If you are in error, you correct whatever error you have so you come out right and exact which is a method of being righteous. Since mathematics is based on logic and reasoning, *living mathematics* (as some of us tend to say) is using logic and reasoning on a daily basis instead of being religiously radical or radically religious.

"And in a while the style'll have much more value/
Classical/too intelligent to be radical/
Masterful/never irrelevant/Mathematical."

—RAKIM ALLAH
"Don't Sweat the Technique,"
off the 1992 album *Don't Sweat The Technique*)

Being mathematical on a daily basis requires practice of orderliness of thought and behavior. Since mathematics is about order, we too must be about order. Some may wonder if there is any relationship at all between Supreme Mathematics and the mathematics some of us grew up dodging and avoiding. The answer is yes, and here are four good reasons: 1. The definition of mathematics in both explicitly and implicitly mean knowledge, study and learning; 2. The only difference in the two is the word *Supreme* that implies a different method of application; 3. Both involve the study and application of quantity (numbers), structure, space and change; 4. Both are used together with alphabets to solve problems. It is true that all 26 letters of the alphabets are used in mathematics. Galileo was spot on when he said: *"The universe cannot be read until we have learned the language and become familiar with the characters in which it is written. It is written in mathematical language, and the letters are triangles, circles and other geometrical figures, without which means it is humanly impossible to comprehend a single word. Without these, one is wandering about in a dark labyrinth."** Therefore, the daily application of Supreme Mathematics and Supreme Alphabets as they are called in our culture are fundamental to the growth of our knowledge, study and learning.

---

* The Assayer, *Opere Il Saggiatore* [in Italian], p. 171

The problems that we face from time to time in life can be solved when we apply our mathematical minds to them. This is doing things and living according to order and order is the basis of peace. When things are out of order, what do you naturally want to do? You want to restore order so you can live in what? Peace. If you are the type of person that tends to want fast and easy solutions to problems you don't know or understand the answers to, consider this. In mathematics, something unknown is equal to (X). We can't solve for (X) until we reason through the problem. Once we break down the variables of the problem, we can then solve for (X). This means the more we know, the better we will be able to solve whatever problems are in our way. People may offer an answer before knowing the all the variables in which case you will end up with the wrong solution and the problem will still exist and will probably be more confusing. We don't want to settle for any answer, we want the answer that's correct, so we'll have to ask the right questions to find the right answers. Sometimes we waste time asking and debating the wrong questions which keeps the problem from being solved.

We also have a saying that we speak mathematics and mathematics is our language. This is not solely based on our unique way of speaking. It also applies to our everyday speech. We can translate phrases into expressions and sentences into equations, combining those skills to communicate and solve verbal problems. People do this without even realizing it.

"Mathematics allows for no hypocrisy and no vagueness."
—STENDHAL*

Mathematics, doesn't allow you to make up stuff, invent fanciful stories, or razzle and dazzle people with revisionist history. You simply have to show and prove. No fancy justifications, hypocrisy, vagueness, equivocations or excuses are allowed. It's up to those of us who know the formulas to work it out and simplify our solutions. Hypocrisy is the claim or pretense of holding beliefs, feelings, standards, qualities, opinions, behaviors, virtues, motivations, or other characteristics that a person doesn't in fact hold. Vagueness is the state or condition of lacking clarity, something not definitely established, seen or proven. Stagnation is a state that is contrary to the nature of Islam (Mathematics). Stagnation means to fail to develop, progress, or make necessary changes. It also means to stop flowing. Mathematics is used to resolve the truth or falsity of conjectures by mathematical proof.

History shows us that all great cultures devoted substantial energy to mathematics and to science. Without a sense of how quantitative concepts are used in art, architecture, and science, you cannot fully appreciate the incredible achievements of the Mayans in Central America, the builders of the great city of Zimbabwe, the ancient Egyptians, the early Polynesian sailors, and countless others. Therefore, in this culture that's rooted

* Marie-Henri Beyle (23 January 1783 – 23 March 1842), better known by his pen name Stendhal, was a 19th-century French writer. Known for his acute analysis of his characters' psychology, he is considered one of the earliest and foremost practitioners of 'realism' (a concept in the arts that dealt with the representation of subjects truthfully and accurately, without artificiality and avoiding artistic conventions, implausible, exotic and supernatural elements (in other words, without freestyling or making stuff up which some people tend to do)

in mathematics, our energy should be devoted to developing our ability to reason with quantitative information and critical thinking needed to build.

The dumbing down of the masses has been a widespread epidemic for some time now. Just think about it. A great number of people do not know how to read or do math, but love *entertainment.* One of the reasons for this is that reading and math is not taught in an interesting or fun way. Statistically, many people are functional illiterates and have a sixth to an eighth grade reading level. So if anything is present to elevate the people, the intellectuals and conscious people have to "keep it simple" and "dumb it down."

There are nearly 32 million functionally illiterate adults (or a much larger number) in the U.S. that can't read past an eighth grade reading level. The reality gets grim for younger people in this condition as illustrated in a profound documentary by Rahiem Shabazz entitled *'Elementary Genocide.'* In my opinion, all children and adults should know how to read. This is why I choose to volunteer with the Literacy Volunteers of America years ago. Reading and mathematics are two very important exercises for the enhancement of the brain. The more we do, the smarter we become and the more problems we can solve.

A lot of people dislike mathematics and find it difficult, but a key component to solving our financial problems can be found with the proper use of mathematics. Along with the teachings of knowledge of self, mathematics were heavily incorporated as part of our teachings. Allah taught us Islam is a natural way of life to us, not a religion, so this left us to deal with things on both scientific and mathematical levels. Often called the language of the universe, mathematics is fundamental to our understanding of the world and, as such, is vitally important in a modern society such as ours. There is no line separating pure and applied mathematics, and practical applications for what began as pure mathematics are found all the time.

> *"All science requires mathematics. The knowledge of mathematical things is almost innate in us. This is the easiest of sciences, a fact which is obvious in that no one's brain rejects it; for laymen and people who are utterly illiterate know how to count and reckon."*
>
> —ROGER BACON

When you're referring to addition, subtraction, multiplication and division, the proper word is "arithmetic." Math, meanwhile, is reserved for problems involving signs, symbols and proofs — algebra, calculus, geometry and trigonometry. These processes can be found in our SM and SA depending on your perspective. For example, in algebra, $X + 2 = 5$ the letter x is unknown, but the law of inverses can be used to discover its value: $5 - 2 = 3$, therefore $X = 3$. The letter X is found in our Supreme Alphabets and is used to represent the unknown variable. For example, the letter X was used by the Nation of Islam to replace their slave names, for example, Malcolm X. The X was used in their name because their real family names were unknown.

I remember meeting a brother who substituted *'borning'* Supreme Mathematics for a solid education in the field of mathematics. All he did was born everything out, a bit strange, I might add. Every word he randomly choose, he would find its numerical equivalent and give it some arbitrary meaning. I recall being taught the purpose of that was to deal with our names, not every word in the English language. While the application of our

Supreme Mathematics is central to our understanding and use of the language to each other, the use of applied mathematics should never be denied because it is as important to be successful in life. Think about it: How can anyone be *the Maker* and *the Owner* of anything today without a command of numbers? Emblem, a god who teaches others it is possible to apply ourselves to higher forms of mathematics, strives to expand our application of Supreme Mathematics in this regard.

INTERVIVEW WITH EMBLEM

Starmel Allah (SA): Peace Emblem. May you tell us what your name means, where you are from, and how you came into the knowledge of self?

Emblem: Peace. My attribute is 'Emblem'. My attribute comes from the born degree in the knowledge to knowledge the culture. I've had Knowledge of Self nearly 24 years, it should be noted that me and my enlightener/educator fell out as the result of an irreconcilable dispute that took place approximately 4 years ago. His name is Life Justice Allah and for 20 years we were very peace. I was born and raised in Power Hill. I was born into the Nation of Islam, my attribute then was William 81x. I attended the University of Islam for first and second grade. My mother Linda 13x left the Nation in 1977 approximately a year after Elijah's death and as I got older gravitating to the teachings of the Five Percent was seemingly a natural progression.

SA: What teaching of the Five Percent attracted you?

Emblem: My mother was teaching me that the Black man was God when I was a baby, b.u.t. the way that she described it is that I was the microcosm of the macrocosm and so I knew as a child that the Black Man was God, and so as a teenager to hear of a Nation solely devoted to that subject matter-it was a 'nobrainer', and I knew that this was for me. The fact that the Nation of Gods and Earths fell along street lines was right up my alley because the Nation of Islam in the early 70's was very much 'street' and very thugged-out, and heavily integral with the Philly Black Mafia and I remember looking up to those cats. I Self Lord And Master with a genius and 'knuckle up' type of vibe really was something that fitted my personality as well as my history. The stories that portrayed the Father as a real 'Street Dude' was magnetic in my eyes. His cool, smart, street current of air attracted me to this Nation.

SA: What kind of curriculum did the University of Islam have while you were there?

Emblem: The University of Islam placed emphasis on language arts, mathematics, social studies, astronomy, and Arabic. My exposure to Arabic as a child is the reason why I can read Quranic Arabic to this day. The study of Astronomy allowed me to have a conceptualization of the vast size of the Universe. My love of language arts and writing is still present and partly responsible for love of the written word, and literature. And I venture to suggest that instilling my young mind with a wonder awe for mathematics is why I gravitated toward the Nation of Gods and Earths as a teen.

SA: Tell us about your background in Mathematics?

Emblem: My background in Mathematics goes as high as Algebra two in High School and College Calculus in which I flunked miserably. Failing that Calculus course greatly affected my approach and perception of Supreme Mathematics as well as my approach to mathematics in general.

SA: What did the University of Islam teach as far as Mathematics?

Emblem: I was only 5 or 6 b.u.t. what I do remember is being drilled heavily on multiplication, division and fractions at a very early age. Elijah died in 1975 and I was only six, however my mother was pushing me at a very early age. I remember the headaches and the seriousness of the situation. Elijah wanted young mathematicians. He understood that mathematics was critical for the elevation of any Nation and he described the few branches of mathematics that the colored man demonstrated mastery over as 'child's play'.

SA: Can you explain what the Order of Operations are and how a person can apply the Order of Operations to Supreme Mathematics.

Emblem: Parenthesis, Exponents, Multiplication, Division, Addition, Subtraction are the essential orders of operation in solving a general algebra problem. How they could possibly affect one's approach to knowledge wisdom cipher, Supreme Alphabet and living out Supreme Mathematics is of great importance in my mind. Presently, many in our culture add quantitative values together when dealing with today's degree. My position is that adhering to the proper rules of math could indeed elevate our Nation into the prominent academic arena of mathematics. Allah School should be a place where the actual laws of math are emphasized and adhered to. Such adherence to the laws and functions of Math would usher fourth an ever advancing creed of teaching the babies of our nation. Personally, I think teaching babies 'all being born' for the next ten years will hurt the babies more than help them. In order to keep up with this ever advancing world it is imperative we prepare ourselves, particularly our babies for the Wild Style of the future soon to come.

SA: Can the higher applications of mathematics really be used in conjunction with Supreme Mathematics, if so, how?

Emblem: Using Supreme Mathematics in conjunction with mathematics is necessary for the advancement of the Nation. Today's Mathematics is Understanding Cipher, in the one to forties it goes into germs: Black and Brown. Germs points to Deoxyribonucleic Acid, a double helix chain. How do you approach the cutting edge technological study of these so-called germs with poor math scores? You don't, you get left behind and though you may discuss these germs, theorize about these germs on some street corner or park somewhere in a casual fashion does not constitute as supreme expertise. Where is your microscope, or your data from other scientist around the world who play with the dominant and recessive characteristics of genetics daily? What are their findings and how do they relate to your own? These scientists are manipulating germs/genes to the point in which they are literally making designer genes. Will our babies ever be allowed to stand firmly and competently on that ledge of knowledge? They will be inept to deal with such technical understanding if they are not schooled in the traditional text and rules of quantitative math which is the price of admission to get into the high sciences being dealt with by other scientist.

SA: We have a common saying, "Knowledge is infinite," however many may have a hard time grasping the perspective you have, especially, if one is not really dealing in real mathematics. Some people don't even like mathematics. So, how would you show and prove this knowledge you are giving is right and exact?

Emblem: That which proves that what I am building on is right and exact is the world around us, those who may reject my claims all have a smart phone I'm sure. Put ten Gods in a room and ten Asian engineering students in a room and give them the assignment to build and design a smart phone from the individual components needed. Which team will be victorious? The team with a background in the hard sciences will win. 'All being born' works just fine for casual philosophical banter, b.u.t. in all actuality you can't build an actual 'thing' with it. Yeah, you can keep your family together with it, b.u.t. Muslims keep their family together. The cultural aspect of family can be held in tact, b.u.t. to negate the reality of the real world and the math that this world is about is just as bad as believing in a mystery God. Planes, computers, satellites, you won't be able to be a good criminal in the next 50 years without computer skills and high mathematical aptitude.

SA: When we say all being born we are combining two numbers that are inter-related to bring about an understanding of reality as it relates to the Arabic numbers 0-9. Can you elaborate a little further on why should Algebra be used to manifest reality? Shouldn't we just keep it simple?

Emblem: The number '1' is inter-related to the next 6 numbers that will be drawn in the Mega Millions, or Powerball Lottery. My social security number is inter-related with a random phone number on any page of any phone book, meaning all numbers are inter-related to infinity. The inter-relation is essentially arbitrary and random. Every number is inter-related to any other number. All numbers are related. You can't walk up to a house and say the address is '34 Such and Such Drive' is all being Born to God. You could, however the real question is what does such an answer yield in such an arbitrary setting? Coincidence within the inter-relationships of numbers will only yield some good principles to build upon, meaning you and the next God can go over some degrees and what not. And that's cool, nothing is wrong with allowing 'All Being Born" to reinforce the principles in Supreme Math, Supreme Alphabet and 120. Outside of the existential, arcane and esoteric reality of the Asiatic world another world awaits and it's the world that you live in daily. Your thoughts dwell in the Asiatic world, yet nothing you come in contact with daily acknowledges or confirms the Asiatic world. How computationally accurate could an individual be in the real world if he or she dealt with all being born? No one, not a one of us deal with all being born in reality. If I pay $1100 dollars a month for mortgage and I sent my mortgage company 2 dollars because I told them I digit summed the amount I would soon be out of doors. This is the most practical example I can give. I dare any God or Earth to 'All being born' their bills. The fact is we have enough common sense not to deal with all being born in the real world. We all know better. We all know there is no mystery God so we don't deal with a mystery God. Why? Because the Mystery God is not real! So why should we deal with a pseudo mathematical system that you could ill afford to use if you were constructing an actual building? Would you send your child to a school and the building was not constructed with real mathematics? Where the roof on the building was supposed to be 42 square feet by 42 square feet b.u.t. the all being born construction crew made the roof 6ft by 6ft? Or would you get on a plane that required 1000 gallons of fuel to fly across the Atlantic and yet the Asiatic Airline put just 1 gallon in the plane? That plane would go down real quick. One could argue that we are too wise to do such a thing. We are indeed, so why would we discount the place value of a unit? The number 20 represents wisdom (or two) units of ten, the cipher is in a place that represents 'ones'. To collapse place value blindly is not a coherent mathematical system. A mathematical system has no contradictions, so thus to contradict or discount the value of place is to refuse to acknowledge that which is actually there. Actual facts deal with size and dimensionality. We don't collapse the dimensions of the Pacific Ocean to a singular number. We acknowledge the full place value: 68,634,000 square miles. If today's date is the 30th I can't go back in time and make it the 3rd. To do so takes me out of time and thus out of reality. Those that want to deal with All Being Born be my guest. They have that right, b.u.t. none of us have the right to expect our babies to compete with the other civilizations of the Earth and we the adults in the Nation of Gods and Earths have imparted to the babies 'All being Born' along with the Math I.Q. of Corky the Down Syndrome special needs child. I gravitated to this Math because I thought the mathematics would be such that I could teach Calculus or Trig easier. Such is not the case. So many people are fighting me on this issue. Why? Because movafuckuz are having traumatic flashbacks to when they flunked pre-Algebra in the eighth grade and Now here comes this nigga Emblem trying to tell me I'm not really a math genius.

3x is an Algebraic expression that combines Math and Alphabet. Understanding the Unknown. Within every 24 hour cycle we are dealing with the unknown, mindfully we go into the unknown with a determined idea b.u.t. unknowns creep in daily, things we didn't predict. We draw upon 120 for answers to solve 240 which is the daily reality. Every day you get hit with a big ass Algebra equation, meaning everyday you're dealing with one problem after the next. 120 is great, it provides a wealth of answers within a given science of life framework. What does it mean to be an engineer? How could you be an Engineer, or a Scientist in the real world without a solid math background? Would I want my baby to really keep it simple? Hell no and emphatically no! I want my seed to be on equal footing with those Chinese babies in the East. All being born is good to review principles and lessons within this math however you can use Algebra to review deeper and way more complex scenarios within the math itself. When I signed up for this math I did so with the intent of becoming a scientist as well as a mathematician. I will become that and hopefully our babies will as well. Peace.

SA: Peace Emblem.

The lesson that should be taken from this is there are no limitations when dealing with Supreme Mathematics and Supreme Alphabets. They are supreme in that we can take them to the highest level based on our application of them. Mathematics is the language of the universe and therefore, the application of mathematics is infinite. The only limitations we have are the ones we impose on ourselves. It will be extremely helpful that a person is as versed in mathematics as in the Supreme Mathematics they are studying because the relationship between the two is inseparable. Mathematics is about applying methods and operations to finding solutions to basic and complex problems.

The methods of adding, subtracting, multiplying and dividing are integral tools to help us find "correct or proper" solutions not "temporary" solutions. Some of our problems are approached in ways that are not mathematical (i.e., logical or reasonable) but we expect a logical or reasonable solution. Oftentimes, this happens because our understanding of the problem is incorrect. A reading of a problem must be correct in the beginning in order for us to know what to do to come out with a proper answer or solution. If a problem is an economic, the answer or solution must be economic, not emotional or political. But, sometimes people give emotional responses to economic problems.

Therefore, being mathematical (Knowledge, Wisdom, Understanding, etc.) is necessary on all levels. Looking to our Supreme Alphabets is no different as we must find (X = the Unknown) to be able to answer or solve for (Y = Why) things happen. When we do find the answers, we can (Z = Zig, Zag, Zig) through our problems without wasting much time.

CHAPTER 13

Do We Use 10% of Our Brain?

THIS IS A common misunderstanding and the 10% actually have a lot of people fooled about their selves. *"He keeps them blind to themselves so he can master them."** The evidence says the average person uses 100 percent of their brain every day, but here we want to explore how people tend to use about 10 percent

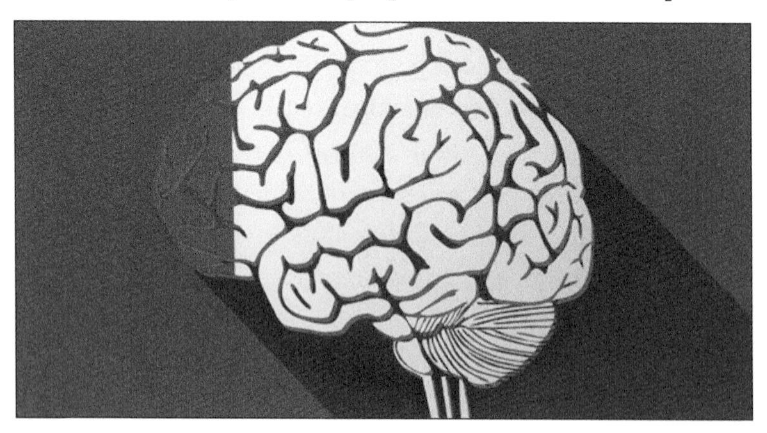

* *6th Lesson, 1-14*

of *what we know.* In the previous edition of *The Righteous Way,* I briefly explained using our minds to achieve success is a matter of how much we make of what we know and how much we apply ourselves. After you have read this book, I would only ask that you try to apply 100 percent of what you have learned in this book, the previous edition, or any other book you may have read before or plan to read in the future. You'll see the difference applying yourself 100% makes.

The human brain is complex. Along with performing millions of mundane acts, it composes concertos, issues manifestos and comes up with elegant solutions to equations. It's the wellspring of all human emotions, behaviors, experiences as well as the repository of memory and self-awareness. So it's no surprise that the brain remains a mystery to some people. Adding to that mystery is the contention that humans "only" employ 10 percent of their brain. If only regular folk could tap that other 90 percent, they too could become savants who remember π to the twenty-thousandth decimal place, have telekinetic powers, and build pyramids with their minds without tools or a labor force.

Though an alluring idea, the *"10 percent myth"* is so wrong it is almost laughable, says neurologist Barry Gordon at Johns Hopkins School of Medicine in Baltimore. Although there's no definitive culprit to pin the blame on for starting this legend, the notion has been linked to the American psychologist and author William James, who argued in *The Energies of Men* that "We are making use of only a small part of our possible mental and physical resources." He didn't say we use 10% of our brains. It's also been associated with Albert Einstein, whose autopsy suggested the small portion of dark grey matter was the portion he used most. This kind of stuff makes for entertaining movies like *Lucy* that take hold of our imagination.

The myth's durability, Gordon says, stems from people's conceptions about their own brains: they see their own shortcomings as evidence of the existence of untapped gray matter. This is a false assumption. What is correct, however, is that at certain moments in anyone's life, such as when we are simply at rest and thinking, we may be using only 10 percent of our brains.

"It turns out though, that we use virtually every part of the brain and that [most of] the brain is active almost all the time," Gordon adds. "Let's put it this way: the brain represents three percent of the body's weight and uses 20 percent of the body's energy."

The average human brain weighs about three pounds and comprises the hefty cerebrum, which is the largest portion and performs all higher cognitive functions; the cerebellum, responsible for motor functions, such as the coordination of movement and balance; and the brain stem, dedicated to involuntary functions like breathing. The majority of the energy consumed by the brain powers the rapid firing of millions of neurons communicating with each other. Scientists think it is such neuronal firing and connecting that gives rise to all of the brain's higher functions. The rest of its energy is used for controlling other activities—both unconscious activities, such as heart rate, and conscious ones, such as driving a car.

Although it's true that at any given moment all of the brain's regions are not concurrently firing, brain researchers using imaging technology have shown that, like the body's muscles, most are continually active over a 24-hour period. "Evidence would show over a day you use 100 percent of the brain," says John Henley, a neurologist at the Mayo Clinic in Rochester, Minn. Even in sleep, areas such as the frontal cortex, which controls things like higher level thinking and self-awareness, or the somatosensory areas, which help people sense their surroundings, are active, Henley explains.

Take the simple act of pouring coffee in the morning: In walking toward the coffeepot, reaching for it, pouring the brew into the mug, even leaving extra room for cream, the occipital and parietal lobes, motor sensory and sensory motor cortices, basal ganglia, cerebellum and frontal lobes all activate. A lightning storm of neuronal activity occurs almost across the entire brain in the time span of a few seconds. The cerebrum, cerebellum, medulla oblongata, and brain stem are all used. That's 100%, not 10%.

"This isn't to say that if the brain were damaged that you wouldn't be able to perform daily duties," Henley continues. "There are people who have injured their brains or had parts of it removed who still live fairly normal lives, but that is because the brain has a way of compensating and making sure that what's left takes over the activity."

Being able to map the brain's various regions and functions is part and parcel of understanding the possible side effects should a given region begin to fail. Experts know that neurons that perform similar functions tend to cluster together. For example, neurons that control the thumb's movement are arranged next to those that control the forefinger. Thus, when undertaking brain surgery, neurosurgeons carefully avoid neural clusters related to vision, hearing and movement, enabling the brain to retain as many of its functions as possible.

What's not understood is how clusters of neurons from the diverse regions of the brain collaborate to form consciousness. So far, there's no evidence that there is one site for consciousness, which leads experts to believe that it is truly a collective neural effort. Another mystery hidden within our crinkled cortices is that out of all the brain cells, only 10 percent are neurons; the other 90 percent are glial cells, which encapsulate and support neurons, but whose function remains largely unknown. Ultimately, it's not that we use 10 percent of our brains; rather, we merely only understand about 10 percent of how it functions.

Some real smart folks at Harvard University did a study and determined that we spend about fifty percent of the time lost in our thoughts. That means about fifty percent of the time we are not using our most powerful tool because we are distracted by what's in the future, and held up by what's in the past. Sometimes we say, *"Aw, I'm not going to make it,"* that's five or ten minutes from now. We say, *"I don't like the way he/she talked to me."* That was five or ten minutes ago. What we need to be doing to be able to use the most powerful tool is we need to be in the *right now*! Let's stop dwelling in the past, doubting ourselves and worrying about the future, the time is now! If we want to change our situation, it starts right now! If we want to take control of our lives, it starts right now! If we want to improve ourselves, it starts today — right now! The world is divided by those who know and those who do not know. Those who do not know will always be subject to those who know. As stated earlier, the bloodsuckers of the poor (10%) take advantage of this and take advantage of anyone who doesn't have or know how to use their Knowledge of Self. There will never be equal treatment between those who *use more of knowledge* and those who *use less of knowledge.* Whenever knowledge or information is limited, concealed or denied from people, the concealers have an advantage. Most people use all of their brains, but some are privy to more *knowledge* while others are left straggling behind with outdated or irrelevant knowledge. We must privy ourselves to knowledge that should be applied one hundred percent in order to survive, thrive, be efficiently competitive, and successful in all areas of civilized life. We are not dumb, or inferior to anyone on the planet. *We simply have to learn to go get the knowledge we need and then* ***USE 100% OF WHAT WE KNOW!***

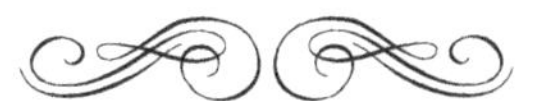

"Sun, know you are Allah, never deny yourself of being Allah, even if the whole world denies you, never deny yourself because it's your own doubt that can stop you from being Allah."

—ALLAH

CHAPTER 14

"Renew" Instead of "Repeat"

OVER THE LAST 50 years, we have learned a lot about the dynamics of movements and what can happen inside movements, and the various social, economic and political conflicts that can come up. Leaders came and went, some were sincere and others were not. Some people were killed as others acted out of rage on impulse and loyalty to other people having opposing views. The decisions made on one level changed how we dealt with each other on another level. The domino effect was that it was more than one word that could change a nation. One person's actions could change the nation as well, and it did. One person's words or actions in one area could have a lasting effect on the minds of many without them even realizing it. Someone can do something today that can change the

course of history, so we want to renew history not repeat history. To renew means to make something, someone, or someplace new again. Something can't be made new again by repeating the same thing. When you do what you always did, you'll get what you always got; and if someone does the same thing over and over but expects a different result each time, they might need a shrink.

The time it takes for the Earth to rotate completely around on its axis is what we call a day. It's Earth's rotation that gives what appears to us as night and day. One complete rotation actually takes 23 hours, 56 minutes and 4 seconds. So science and nature teaches us that no two days are exactly the same because the Earth is always turning. But people forget these little facts sometimes as they live day to day. We tend to forget that we are living in the present at all times. We cannot change what happened the day before any more than we can change what happened 50 or 100 or 1,000 years ago. What we can do is predict, plan and make choices based on the lessons we learned about the past. If we don't learn our lesson, we are doomed to repeat them again. So in our lessons, it states that we make history to equal our home circumference. Every day we live *we make history.* We make history. We shouldn't repeat history.

If we make a mistake today, it is okay, because you can learn from that mistake and correct it. Tomorrow, you will be a wiser person and you can teach someone a valuable lesson based on what you know from your experience. You can help students develop theirselves personally and networks with peers. You can be a non-judgmental listener to a brother or sister and help guide them out of ignorance, weakness or self-destruction. You can coach and teach students how to be successful anywhere they go. But you can't teach what you don't know. So we need you to be successful so others seeking to follow someone can follow you. That's what righteousness is all about. Our planet doesn't travel backwards neither should we. Going backwards or doing things based on yesterday's circumstances with yesterday's mentality and yesterday's beliefs would be as wrong as thinking the Sun revolves around the Earth. Some are stuck in a particular point of time, whether it's the 50s, 60s, 70s, 80s or 90s. They want to relive the good old days but those days came and went. As a result, many people have not been able to be current in their thinking or beliefs. That is detrimental to that person and all other people involved with that person because their outdated ideas could be no longer relevant to anyone.

We renew our history not only through going over lessons, but by teaching the young to be participators in their own future. We have to advocate for students and teach them how to advocate for their selves. If you think it's all about you, then everything will end with you. Many smart phones automatically update themselves with new versions of software when the previous software becomes obsolete. Imagine if we did that every year to adapt to the new challenges and tasks at hand. People have to keep current with knowledge and information. Decisions have to be based on current and correct facts, not outdated information and beliefs. The person having the most proper knowledge (the best knower) is best qualified to lead and correct those with the least knowledge. So everyone under these studies learn to renew their history and their knowledge by constantly studying. Even at work, it is common for companies to have seasoned employees undergo new training to keep up with new changes.

Our planet rotating every day is proof that no two days are the same. So nothing stays the same forever. Everything changes and sometimes it happens in subtle ways and sometimes in very obvious ways. Seasons change from spring to summer to autumn to winter and back to spring again. Everything is comprised of atoms that are always moving (some very fast and some very slow). When the speed of atoms changes in a thing (due to a change in temperature or pressure) the composition of that thing also changes. Change is inevitable and it's an undeniable property of our universe. So people who resist change are only struggling against the very nature of the universe. It's like trying to escape gravity on Earth with weights around your waist.

The most knowledgeable people anticipate and prepare for change. They become swift and changeable in thinking. This way, they are flexible enough to adapt to the change yet continuing to be who they are. Since we live in an ever changing universe, it would be wise to always be ready for change. Change is not always a bad thing. Change can be a good thing, too, depending on how current and correct your knowledge is.

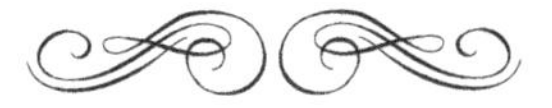

"As life change you change. If you don't change you are going to die. Change with the young people. If you stay around young people you'll stay young."

—ALLAH

CHAPTER 15

Constant Elevation

A COMMON PHRASE USED by Five Percenters is that we must deal with *"constant elevation."* This phrase made a significant impression on me as a teenager because those words provided motivation for self-improvement and personal transformation. I internalized those two words: **constant elevation**. They say a

word can change a nation, so just image what two words can do? The word constant means something happening all the time no matter what. Mathematically, constant means to be unchanging. Elevation, in this context, is to ascend or go up in status as an indication of improvement. So improvement is always happening and should be moving you upward. I think there is something significantly useful here for some of us if we find ourselves feeling stagnant or down.

Constant elevation is movement in a positive direction. Imagine a graph where positive movement on the Y-axis is moving upward and positive movement on the X-axis is moving to the right. This means when you're dealing with positivity, you'll be *moving up* and *in the right direction.* If someone is moving to the left or downward, yet they say they are dealing with positive movement (or constant elevation), they would not be correct. From a mathematical perspective, our teachings are leading us in this positive direction to elevate us or get us moving in the right direction. Every day, we should be *stepping up* our effort to improve ourselves. Remaining the same, on the other hand, defeats the purpose of why we teach and learn. These teachings elevate people, picking them up from a wretched state and standing them up right in a moral perpendicular. For example, when a student takes on a new name (what some mistakenly call an attribute), some change will occur in that person's view and attitude about themselves. They will surely experience (if serious) a slight degree of transcendence in understanding. If a person goes from being Kenny to Wise King, a subtle shift in thought process, perspective, attitude, or behavior will occur. Usually, by the time a student learns the principles (or laws) of Supreme Mathematics and the

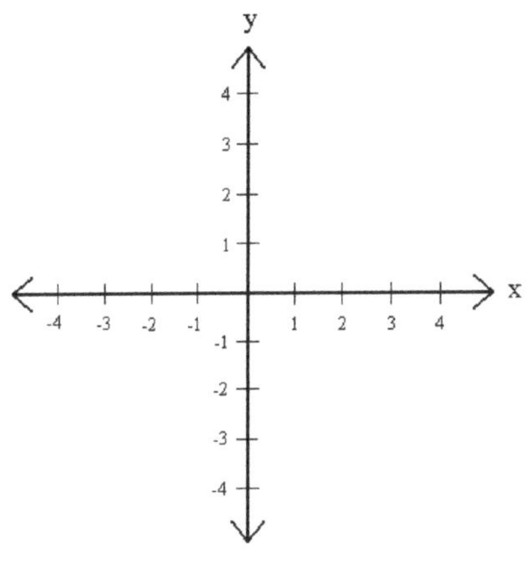

basic letters of the Supreme Alphabet which make up our language, they would have eschewed pork and changed their diet in some kind of way, studied the history of Allah, and established an ongoing dialogue with their enlightener or educator.

Constant elevation, as a positive act of acquiring knowledge, takes your mind through a refinement process. Through knowledge, we acquire power (by learning truth) to refine ourselves mentally and morally. Some people refuse to deal with this process. Some will exhibit power with no refinement. Some don't even care about getting an education. They want to teach people to clean their selves up but won't do the same for themselves first. They want the power but not the refinement. Remember, in order to change the world, you have to first change yourself. The Power of a Cipher is determined by everything from the Knowledge to the Culture of that Cipher. *Power* is the ability to speak or do something based on the truth you know, but *refinement* is to say or do it in a righteous (or well-mannered) way. Refinement is a process everyone must go through in each cipher to clean their selves up or to improve (whether the word *'refinement'* was used in a lesson or not). The brain is designed to constantly learn to make us better. Who doesn't want to think, feel, speak and act better? Whether we admit it or not, there's always something we can do better. The question is whether we are willing to be constant in our pursuit of that improvement?

Gods from Far Rockaway, Queens, Early 1980s,
Courtesy of Jamel Shabazz & God Barsha Allah

Gods from Far Rockaway, Queens (1983)
Courtesy of God Barsha Allah

CHAPTER 16

Change What? Why Change?

GOD SHAMMGOD PLAYED in the NBA with the Washington Wizards during 1997–98 after being drafted by them in the 2nd round (17th pick) of the 1997 NBA draft. In high school, he was known as Shammgod Wells. However, upon attending Providence College, he was forced to either legally change his name to Shammgod Wells, or to use his legal name of God Shammgod. Because he did not have the $600 required to legally change his name, he was known as God Shammgod from that point onward.*

For Shammgod's father, seeing his son's journey has been rewarding, and it has been a process he and I can relate to. He never returned to prison after his release in 1980. He had spent nearly four years in prison after his son was born before he turned his life around, becoming a more involved parent and graduating from college. God Shammgod knew the value of maturity and

* Weber, Jim. *"God Shammgod's unforgettable name is still bringing him fame – The Dagger – NCAAB Blog – Yahoo! Sports"*

education. He earned a bachelor's degree from the New School for Social Research in New York and worked as a central office technician for Verizon for 27 years before retiring in 2008. He also spent time as a volunteer.

As his son grew up and became closer to him, he used his own story as a message that it was never too late to make changes in life. *"Whether my son was overseas or wherever he was, I always told Shamm, 'There's more to life than New York City,'"* Shammgod Sr. said. *"Most of the time when Shamm came back to New York,*

God Shammgod (r) Ben Solomon for The New York Times

Carmelo Anthony and the Universal Flag

he would go around so-called homeboys and stuff. He saw what I was saying to him, that you keep coming back to the same thing since you were a child. Why don't you go someplace and reinvent yourself? Because you can't do it here in New York. It's too much of a distraction. I always told him: 'I don't care where you land at. Just go and stay and do something with yourself.'" After playing in China, Poland and Saudi Arbaia, Shammgod is now again a student at Providence College, but now he's an undergraduate assistant for the men's basketball team, which won last year's tournament.*

This goes to show in life, you can change course and do something new. Sometimes it's your environment that has to change. Sometimes, we have to change the type of people we hang around, or maybe, it's our attitude. But if you want to change anything, you have to change the type of thoughts you have. It is our thoughts that change everything. Everything around us changes. According to philosopher, Heraclitus, the only constant is change. As we enter a new cycle of time, we are naturally drawn into a new cipher. According to our mathematical ideology, we are now in the time of knowing, speaking, understanding, and living according to the power of the cipher. That power (whether individually or collectively) must undergo several stages or levels before the next cycle. More people will be tapping into their personal power in order to survive and more people will join forces to enhance their collective power. Others will forge new alliances after learning from the failures of past alliances. This will be taking place because all things must be refined or improved in

---

* Tim Casey, *"Known for a Dribble, God Shammgod Crosses Over to Teaching,"* March 11, 2015; www.nytimes.com/2015/03/12/sports/ncaabasketball/god-shammgods-latest-move-at-providence-teaching-as-he-learns.

order to continue growing, developing and surviving. This is a law of evolution, and by evolution I am specifically referring to *change over time.*

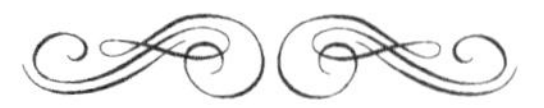

"For the benefit of any country certain people got to change. It's not a shame in change."

—ALLAH

ALTITUDE ABOVE ATTITUDE

What is altitude? What is attitude? Why must your altitude be above your attitude or someone else's attitude? *Did you know that your success can just be an* ***attitude*** *or* ***altitude*** *adjustment away? Are you optimistic or pessimistic? Can you change your attitude if you need to? How high are you're expectations or aspirations? Are you thinking big enough? How much effort are you putting in? Can you go further? How about farther?* Altitude or height is a point of elevation usually in the vertical or "up" direction above the ground or sea level. This is used here to refer to a person's state of mind and level of intelligence. Attitude is a way of thinking or feeling about someone or something, typically one that is reflected in a person's behavior. Your attitude can be good or bad but it is usually formed on *your mood* at the time about someone or something.

Both altitude and attitude are changeable and can be shaped and formed from a person's past or present. According to psychology, attitudes vary from extremely negative to extremely positive, but people can also be conflicted or ambivalent towards someone or something meaning that they might at different times express both positive and negative attitudes towards the same person or thing. This explains why some people can love and hate someone or something at the same time or at different times. Sometimes, we may even get confused or be conflicted about people and situations. Putting knowledge first helps to avoid these types of confusion and conflict which lead to different attitudes. A good way of rising above this confusion is to do the opposite of what contributes to confusion:

C - Clarity in thinking is absent
O - Observation is not consistent
N - Neutral mind set is absent
F - False hopes and expectations are put before knowledge
U - Unawareness of ideas that undermine our teachings
S - Strategic decision making is absent
I - Impulsive decisions are predominant
O - Our judgments are rushed
N - Negative views reach us faster than positive views

We don't want to keep having the wrong attitude about the right thing or the right attitude about the wrong things. That hasn't gotten us very far. And we don't want to have the right altitude (or state of mind) with the wrong attitude, or the wrong altitude with the right attitude (not the right state of mind but having good intentions). The goal here is to increase our altitude and have the right attitude about the right thing. Our teaching of proper education always causes elevation in the minds of new students. In turn, elevated minds nurture the right attitudes needed to accomplish any task or goal at hand.

An observation of growing social media trends will show there is more "building" happening in the form of Internet interaction than the work that is so needed in communities across the country and in the world. Studies show more and more people spend more time on social media each year. When we look at the staggering numbers of purported Gods & Earths networks and groups, one has to ask in all fairness, is the addiction to technology helping or hurting us as a people? Every day, people turn to social media with questions, concerns, issues, and problems, but get back many opinions. The issues we had before social media still exist now. So why not use social media to start solving some issues?

Some people are afraid that "big brother" is watching and will disrupt efforts, so what happens? Many people will prefer to use social media for anything but nation building because it is too serious a subject to discuss openly. We are made to yet again live in fear of being accountable and responsible within our own culture. Social media then becomes a dumping ground for our ideas, thoughts, and feelings with little to no action taking place outside the matrix. Moreover, how do we filter truth from falsehood with all the bombarding of information? How can we be certain the information we are receiving is accurate? One good method is by relying on media outlets with a track record of providing solid information and adhering to journalistic standards. Another is to check the sources independently. Today, the Internet has made it possible to verify some information simply by clicking on a hotlink to the original sources. There is a bunch of inaccurate information out there about Allah's Five Percent Nation of Gods and Earths. More and more people have come to rely on citizen "journalists," increasing the probability that false information will be widely distributed by unscrupulous individuals. Thus we must not take anything or anyone on face value and exercise a high level of caution when reading, listening, and viewing.

Building is an activity not for the fearful or skeptical, but for the mathematical. It is for those who can think clearly, measure precisely, and apply skills to turn ideas into realities. It's not to go endlessly back and forth with someone behind a computer or cellphone. Our definition of building should not include the repeating of misinformation, indoctrination, and unfruitful ideas. It is an act of creating something based on the highest degree of intellect manifested through work.

If we had as many Gods & Earths attend and participate in Parliament as we do on the internet, our nation would be changed forever. Parliament happens once a month and it is the one day reserved for us to come together and make decisions beneficial to the growth and development of our nation. Parliament is the time where nation members can address the nation with their concerns and get a consensus for an idea, but only a small fraction of nation members attend and participate in such parliaments. The one mind and one understanding people talk about is manifested by too few, while too many offer opinions on social networks which is not the best forum for nation concerns. The nation must reform itself in this area and maximize its use of Parliaments. Why? Because if a society or community of people are civilized, that means they have laws they abide by. If this society or group of people adhere to the principles of FREEDOM, JUSTICE AND EQUALITY, every person who abides by these principles must have the freedom to exercise their vote for that which is just, equitable and in the best interests of the majority. Arguments on social networks have failed us in this regard and should not substitute Parliaments. Those who don't agree with our ethos and way of life yet offer opinions and criticism are those with opinions only. Much of this is evidenced by the ratio of online activity versus people offline in the world pooling their resources to build together.

While all people are welcome to our functions, certain rules and guidelines should be in place to avoid confusion and maintain order and peace. In the past, we have had people misrepresent our teachings through the videos or photos they shared. Such videos and photos have done more harm than good to our mission, they depart from our ethos as a culture, and it's our responsibility to take necessary measures to avoid confusion as we move forward.

The laws of respect and etiquette necessary to maintain ORDER and PEACE in any environment or medium of communication between persons claiming to belong to said body are governed by Supreme Mathematics. While many profess to belong to such a body set forth by Allah, it is incomprehensible to imagine anyone who would depart from his basic teachings such as to take nothing on face value and to not talk about other people. People who depart from the basic laws of respect and etiquette due to technology, yet claim to be civilized and belong to a body of wise and civilized people who preceded them brings great shame (and not honor) to the entire body. When the will of such persons are not in alignment with the will of Allah, yet they claim to be of Allah and dare to claim they are Allah, will fall among those who already fell victim to Ezekiel 3:18 and St. Luke 12:47.

Why am I doing all this? Why do I care? It is because my focus is on the issues that are important to our growth and development; creating jobs, owning and developing property, improving infrastructure and making sure that the next generation has an even better shot at life. It's not enough to stand on the sidelines and simply complain about the status quo. I build because I want to change things for the better. I build to make a difference. The purpose of building is to create change, not to keep things the same.

Constant elevation keeps your state of mind above the ground or sea level attitudes of other people who we describe as haters or negative people. Such people seek to only bring you down because they see you elevating. What happens in a barrel full of crabs is what makes that all too familiar saying famous. When a single crab is put into a barrel with no lid, they surely can and will escape. However, when more than one is in the same lidless barrel, none can get out. If one crab elevates itself above all, the others will grab this crab and drag them back down to share the mutual fate of the rest of the group.

The Crab-in-a-barrel syndrome is often used to describe social situations where one person is trying to better their self and others in the community attempt to pull them back down. This can happen anywhere (at work, in school, in business, etc.) it's not just a hood thing. They see you climbing up but they feel they have to pull you down in order for them to climb up. That crab-in-the-barrel mentality should never have a place in our culture. So our altitude must always be above whatever attitude or opinion someone might have of you.

There are people in the NGE and the NOI who don't know their lessons. Furthermore, the same ones who don't know their lessons seem to have the most to say and argue about. When people know and understand their lessons, there's nothing to debate and waste time on that which does not exist. The lessons were made plain so if you don't understand them or can't show and prove, go back and study some more. But, don't waste time going around putting other people down when we all claim to be pulling each other up. That's the kind of hypocrisy that keeps us down, disorderly, dysfunctional, and disorganized. I have seen fruitless arguments spiral out of control to the lowest point of name calling, accusing one another, and even fighting. Regardless to what our differences in view may be, we have got to start looking at our commonalities. The lessons explain to us that it is the devil who *"wants us to think we are all DIFFERENT."* Remember how much fighting and killing took place in the name of religion.

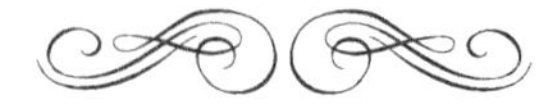

We don't talk about nobody on religion, politics, war and protest in the United States. Because religious people fight against one another. You can't tell me they don't because they do.

—ALLAH

Anywhere on the planet where people start to think they're different you'll see them fighting and killing one another which is the hand of our ancient enemy at work. There is no trace of righteousness where Original people are constantly fighting and killing each other *after* claiming to know lessons. That is a falling from grace. Some of us have family in the NOI and the NOI has family in the NGE. Our NOI family says "As-Salaam-Alaikum" (Peace be unto you) and the NGE family says "Peace." Why is this so hard for us to live out and we both say our WORD IS BOND REGARDLESS TO WHOM OR WHAT. We can have unity without assimilating into each other's way of life. Some of the NGE and NOI can do 100 times better!

We are really supreme beings. Many of us don't understand why or see how because some of us like being average, simple, or trying to fit in with other people so much. We lie, steal, cheat, argue and fight so much we don't realize we fell from grace. Some of us like and even love to lie, steal, and cheat from each other. We look to "get over" on someone or have a warped sense of entitlement. Some of us don't like to pay for anything. It has to be for free. Under those conditions, we can never see the grace we fell from no matter who is trying to show us and teach us a better way. You can forget about social equality or any equality. There's always someone trying to master (manipulate) you under the pretext of being righteous, being your brother, your sister, etc.

There's an analogy in scripture about the devil who was first an angel with God until God commanded his angels to bow to his greatest creation — man. All his angels bowed except for the one who felt he was better than God's greatest creation. As punishment, God cast the now devil out of heaven. Some of us are like that if we can understand the analogy. We fall from grace when we start thinking and feeling we are *better than* people. Think about it. We are on a planet where nearly everyone has a gripe against someone else. There's always something we are fussing about. There's always a war going on. There are those with great attitudes and others whose attitudes are not so great. In all the knowledge we have about civilization (past and present), some of us have yet to understand or really participate in the process of peace because of our attitude. But it's never too late to change that.

CHAPTER 17

Justice Deferred is Justice Denied

BEING A PRISONER is considered by many (including the U.S. Constitution) to have the same status and rights as a slave. When a prison regulation clashes with a prisoner's freedom to embrace God in a cultural way, the courts must strike a balance between the two. Oftentimes, in the lower courts prison officials do not provide any evidence that their regulation serves a legitimate penological interest but simply come up with a post-hoc, speculative reason to justify the restrictive policy. While we have reached settlements in New York and Massachusetts, this rubber-stamping trend has continued in New Jersey, South Carolina, Virginia and Georgia.

The subject of prisoner's rights and treatment of incarcerated NGE adherents was discussed in Chapter 16 of Vol. 1 of *The Righteous Way.* I discussed how prison administrators violated our constitutional, statutory and/or civil rights during incarceration. Since then, our challenge of the New York DOCCS' Protocols reached a settlement on October 29, 2013. In this settlement, myself, God Universal, Magnificent Allah and Born King Allah negotiated the terms upon which would be fair in accordance with our cultural tenets and the policy changes.

It took us five years of litigation and even more years of research and petitions to get us to this point, but we did it. The language of the settlement set forth sincere adherents of the Nation of Gods and Earths ("NGE") shall not only be treated as a belief system entitled to First Amendment protection but shall also be allowed to: hold Civilization Classes, Rallies and Parliaments; the observance of Honor Days (Allah's Physical Birthday on Feb. 22nd, Allah's Show & Prove on June 13th, Allah's Social on Aug. 22nd, and Birth of the NGE on Oct. 10th); wear the Universal Flag and crowns with tassels; including the possession of Supreme Mathematics, Supreme Alphabets, 120 Lessons and other NGE literature and material.*

The settlement was by no means a finalization of the justice we seek. We understood we were fighting for something that was supposed to already be ours — the right to practice our culture. We understood our rights were not given up at prison doors. We weren't merely seeking recognition, we we were seeking the respect of our rights and the enforcement of fair treatment under the law. In society, we still struggle for our rights to be respected.

* Stipulation and Order of Settlement and Discontinuance 9:11-cv-00159 (DNH)-(RFT)

The belief that this culture somehow started in prison is as false as the belief that the world was flat. Although the culture may have had humble beginnings in the streets, it was never intended for any of us to remain in the streets or become a product of the streets. Our teachings' purpose is for us to right wrongs wherever they may be including in the prisons. To not have a sword above their heads, Five Percenters who had so much as a bench warrant turned themselves in order to properly handle their legal issues. Allah aptly stated:

> *"Never let it be said that I taught the children to do wrong. When the knowledge was first born in the wilderness of North America some of my young Five Percenters had to take this knowledge to prison to teach those who had not heard the teachings, these were my first borns. And the only reason why they were imprisoned was for teaching the truth or defending themselves on the street. It is much easier to be righteous in jail than in the street. There is no temptation, no responsibility, they tell you when to get up, when to eat, where you can or cannot go, what you must wear and what you must hear. The true test of a God is in the street showing and proving, taking care of what is his and showing others the proper way and that's merciful because Yacub didn't play those type of games. When one fell victim the penalty was death and was enforced on all who did not hold fast."*

Fifty years ago, police brutality against the Black community was the norm and several incidents (as in the case of the Harlem Six) in the summer of 1964 led to a series of racially-motivated riots in New York, Philadelphia, Chicago, and Jersey City. Fast forward to 2015 and police relations have not changed much at all. Aggressive tones, hostile questioning, excessive force, and the murder of unarmed Blacks are just some of the abuses of authority by police officers toward members the Black community across America.

Today, Blacks continue to give voice to the fact that black lives matter. Of all the people in America, African Americans aged 20-24 are the most likely to killed by law enforcement according to the Centers for Disease Control and Prevention, National Center for Health Statistics. Darrien Hunt, Ezell Ford, Omar Abrego, Tamir E. Rice, Tanisha Anderson, Rumain Brisbon, John Crawford III, Keith Vidal, Kajieme Powell, Akai Gurley, Eric Garner, Mike Brown, Michelle Cusseaux, Jack Jacquez, Jason Harrison, Yvette Smith, Louis Rodriguez, Matthew Pollow, Dontre D. Hamilton, Victor White III, David Latham, Maria Godinez and Sandra Bland are all people of color who were killed by cops in 2014 and as recent as 2015. Some cases are more clear-cut than others, but all of them raises questions about the use of force — like shooting to kill — in policing. They also prompted scrutiny of how cops confront people. These are only a few compared to the many cases (known and unknown) that have taken place since Amadou Diallo and Emitt Till.

In the months since Michael Brown was shot and killed by Officer Darren Wilson, demonstrators have marched in major cities across America, set up street blockades, shut down highways, and faced off against the very police officers they want brought to justice. Grand jury decisions not to indict Wilson, or the NYPD officer who choked Eric Garner to death on film, have reached a tipping point for people exhausted with the number of deaths at the hands of law enforcement. Dozens of congressional staffers walked off of their jobs to the steps of Capitol Hill in Washington D.C. to protest the Grand Jury decisions. This, including 460 years of terrorizing and murdering Black and Brown people, compelled a sea of people to unite and rally with the Honorable Minister Louis Farrakhan for Justice or Else. In support of this cause, many Gods & Earths answered the call and attended the rally on the same day of the anniversary of the birth of the Five Percent Nation of Gods & Earths. As a force of influence within our communities, and as people with knowledge of self, our knowledge was reinforced by the unity of that day. The Justice or Else rally came twenty years after over one million Black men came to the same stretch of lawn between the Capitol and the

Washington Monument to rededicate ourselves to being better fathers, sons and citizens. The messages that were delivered to the masses on October 10, 2015 contained knowledge and truth for the masses to bring back to their communities and homes to make a difference. There were many young people, and many young Five Percenters in particular, who will bring that knowledge back to add on in their communities.

WE
NEWARK

Photography by JP Entertainment

While many are on the front lines and side lines challenging our government for justice, we have another struggle for justice about the conditions in our own community. We owe it to ourselves to be the justice we seek. Justice begins with decency and fair treatment of people regardless of their color, sex, class or what neighborhood they live in. The cornerstone of justice is being just and fair. Therefore, we owe it to ourselves to be just and fair to each other. Every human being is born with rights and if authorities violate those rights (even if such a person broke the law) justice must apply to them as well. When a human being is denied their basic rights because of discrimination or prejudice, there is an injustice there (whether in society or in prison). Although the laws and systems of society may not be entirely perfect, every person deserves their day in court and a fair trial. But the problem the average person faces is being totally or partially unaware of their rights and that is an injustice to your self.

A major issue between law enforcement and Black and Brown peoples in poor communities is the abuse of power and police brutality. Some officers seem to believe they have a license to kill. On the flip side, many people who don't know the law or their rights make themselves easy targets for officers anxiously waiting for the excuse to use excessive force or satisfy a pathological bloodlust with an itchy trigger finger. Of course, all officers are not like this, but the homocidal misconduct of a few, including those officers who keep quiet about these injustices, makes the system appear guilty of murder in the first degree.

Conversely, the police have a job to do as well. They do have to enforce laws, protect property, and save lives. We may have friends and family who work in law enforcement. Some Five Percenters and Muslims are officers. There's a lot of good people out there that need to be protected from those who break the law. Things will only work out when there is a balance of the two sides. That balance is best achieved when the rights of the people are respected and protected. Our ignorance of the law cripples us and before you can stand up for your rights you first have to know them.

On the other side of the equation is we must know the purpose of police and their role in the community. Sir Robert Peel's principles of law enforcement can allow us insight into police practice, but they also show how far some officers have gone beyond the point of reason. These principles give us a criteria from which police behavior can be measured. Known as the father of modern policing, Peel developed the Peelian Principles which defined the ethical requirements police officers must follow to be effective:

1. The basic mission for which police exists is to prevent crime and disorder as an alternative to the repression of crime and disorder by military force and severity of legal punishment.

2. The ability of the police to perform their duties to dependent upon public approval of police existence, actions, behavior and the ability of the police to secure and maintain public respect.

3. The police must secure the willing cooperation of the public in voluntary observance of the law to be able to secure and maintain public respect.

4. The degree of cooperation of the public that can be secured diminishes, proportionately, to the necessity for the use of physical force and compulsion in achieving police objectives.

5. The police seek and preserve public favor, not by catering to public opinion, but by constantly demonstrating absolutely impartial service to the law, in complete independence of policy, and without regard to the justice or injustice of the substance of individual laws; by ready offering of individual service and friendship to all members of society without regard to their race or social standing, by ready exercise of courtesy and friendly good humor; and by ready offering of individual sacrifice in protecting and preserving life.

6. The police should use physical force to the extent necessary to secure observance of the law or to restore order only when the exercise of persuasion, advice and warning minimum degree of physical force which is necessary on any particular occasion for achieving a police objective.

7. The police at all times should maintain a relationship with the public that gives reality to the historic tradition that the police are the public and the public are the police; the police are the only members of the public who are paid to give full-time attention to duties which are incumbent on every citizen in the intent of the community welfare.

8. The police should always direct their actions toward their functions and never appear to usurp the powers of the judiciary by avenging individuals or the state, or authoritatively judging guilt or punishing the guilty.

9. The test of police efficiency is the absence of crime and disorder, not the visible evidence of police action in dealing with them.

From the onset of our 50 year old way of life in America, we have lived differently than the social and religious norms of society. Historically, religion has been used to subjugate, control, and enslave those desiring to be free. All throughout the world, life, language, liberty, and property suddenly became the property of a newly established religion, however, it should be pointed out that no where in any scripture can it be proven beyond a shadow of doubt that God has a religion. As Gods and Earths, we were taught a natural way of life, but it has been a way of life that religious figures have theologically discriminated against.

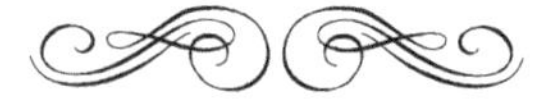

> **"And the Five Percenters, I'm teaching them that they can't go on religion because religion has never done anything for them."**
>
> —ALLAH

THEOLOGICAL DISCRIMINATION

By BORN KING ALLAH

I continue to honor Allah by accepting the responsibility as a Cultural warrior of the NOGE to do the work that is needed to free our people from Theological discrimination. We have not made it, and the battle is certainly not over. Still today the forces that be are doing a most wicked job trying to demonize our most righteous Nation of Gods and Earth's. God's an Earth's pay attention to the call to arms for it will take us coming together as one to defeat the many who seek to destroy our legitimacy. They are attacking us because they don't want the masses to ever accept that the Blackman is God and to make sure the 85% will fail to let us teach them. The power found in keeping the masses blind, deaf and dumb to the reality of God has been used to control people for centuries. As accepting as today's society claims to be, as much as they talk about equality, the NOGE has been denied both acceptance and equality of treatment according to law. This is about to change because as the people most affected by this oppressive bias against us we are putting in the work to change it. The NOGE and its National Office Of Cultural Affairs (NOCA) is working hard and speaking effectively in order to bring relief to our people. We will bring the 2014 modern day persecution we suffer from for choosing the Blackman as God and not the mystery God they offer the masses; to an end. How arrogant is it that the children of our slavemakers who owned, sold, traded, abused, raped, murdered and subjected our children to pedophiles during slavery are today hell bent on telling the NOGE who we are. They are just giving us the name they want us to have and we are supposed to just accept it. The NOGE can show and prove we are the Gods, but still those who oppose us arrogantly continue to call us a gang. This attitude clearly points to a White Supremacist mentality today that permits them to think they know us better than we know ourselves. It's as if they think all they have to do is say something and that's what it is because they said so! This is the same white supremacist

rhetoric of the slavemaker who told Kunta Kinte that your name is Toby because I said so. These people have little respect for God as we understand our reality. To add insult to injury they seek to define Gods who are the highest form of life using the lowest term. They are really attempting to turn the Gods into Gangsters and we see why. They want to destroy the true and living God in order to build more life into their mystery God. They want to destroy the Black God so they can champion every kind of White God they can imagine, Norse God, Greek God, Water God, Fire God etc. According to their White Supremacist rhetoric every God is legitimate except for the Black God who can be seen and heard everywhere. Up until the 1970's Blacks grew up seeing only white images of God and reading books about the heroics of a countless number of other one's. What a distraction, what a great magicians' trick these people have pulled on the world. They get people to deny the true and living God they see every day and to accept a mystery God they will never see until they die. We are tired of America's hypocrisy calling us Black Supremacist for just saying the Blackman is God and at the same time showing white images of God all day, every day! Black people are not physical slaves anymore and some of us are not mental slaves either. There are only a few of us 5% in ratio but that is enough to free the truth from the bondage of your white lies on Black people. We now know how to defeat your liars and lies; we confront you head on with our truth. Take this lie for example "the 5% represent the greatest threat to the social fabric of prisons." A Whiteman Ronald Holvey told it, White America embraced it and created repressive policies against the Gods because of it. But let us examine the liar and the lie. Ronald Holvey is a ex correctional officer out of New Jersey who witnessed us saying the Blackman is God in the prisons he worked at. That's simply who he is so what qualifies him to then make the claim the Gods are the greatest threat? What is the social fabric of prison? So now let's look at the lie. Here is an excerpt from an article called; Fighting Gangs in our Prisons, and in Our Neighborhoods by Ron Holvey Principal Investigator NJDOC. Managing Gangs in the Community: "Currently, seven gangs have been classified as Security Threat Groups by the NJDOC: the Almighty Latin King and Queen Nation, Bloods, Crips, East Coast Aryan Brotherhood, Five Percent Nation, Neta and Prison Brotherhood of Bikers.": In this article Holvey puts our righteous nation side by side with real gangs, you notice there is no mention of the mafia or organized crime figures. Now there is no way they can claim the 5% is the greatest threat to the social fabric of prison, when you look at this list. Really, the 5% is more dangerous than the mafia, organized crime, and real gangs? Based on what, where is his supporting evidence or is it that white privilege means he doesn't need any? This simply does not make sense and this Nation must hold Ron Holvey responsible for making us a target for destruction by multiple law enforcement groups all over the country. This only shows that a so-called gang expert is not qualified to speak on who or what we are. We are not now nor have we ever been a Gang and just because this liar said so doesn't make it so! We will not allow him to assassinate our character and must sue him for defamation! I refuse to allow this dangerous liar to destroy the Nation Almighty God Allah himself built. I can't get pass the fact that organized crime is not a security threat group but the Gods are? The Five Percent, NOGE has a 48 year history that proves we are not a gang. They always point to the actions of individuals with a righteous name who participated in unrighteous activities and make the entire NOGE guilty by association. They don't do that to white people, Italians the Irish or any other group that have real mobs. However they think we will sit idly by while they do it to us. They don't hold the individual acts of pedophilia committed by large numbers of Catholic priests against the church itself. They don't call Catholicism a pedophilic religion. They don't call the Catholic Church a pedophile gang, but let's look at it from someone else's perspective. Here is an article: The Great Catholic Cover-Up The pope's entire career has the stench of evil about it. By Christopher Hitchens: March 15, 2010 "On March 10, the chief exorcist of the Vatican, the Rev. Gabriele Amorth (who has held this demanding post for 25 years), was quoted as saying that "the Devil is at work inside the Vatican," and that "when one speaks of 'the smoke of Satan' in the holy rooms, it is all true—including these latest stories of violence and pedophilia." This can perhaps be taken as confirmation that something horrible has indeed been going on in the holy precincts, though most inquiries show it to have a perfectly good material explanation. Concerning the most recent revelations about the steady complicity of the Vatican in the ongoing—indeed endless—scandal of child rape, a few days later a spokesman for the Holy See made a concession in the guise of a denial. It was clear, said the Rev. Federico

Lombardi, that an attempt was being made "to find elements to involve the Holy Father personally in issues of abuse." He stupidly went on to say that "those efforts have failed." He was wrong twice. In the first place, nobody has had to strive to find such evidence: It has surfaced, as it was bound to do. In the second place, this extension of the awful scandal to the topmost level of the Roman Catholic Church is a process that has only just begun. Yet it became in a sense inevitable when the College of Cardinals elected, as the vicar of Christ on Earth, the man chiefly responsible for the original cover-up. (One of the sanctified voters in that "election" was Cardinal Bernard Law of Boston, a man who had already found the jurisdiction of Massachusetts a bit too warm for his liking.) Very much more serious is the role of Joseph Ratzinger, before the church decided to make him supreme leader, in obstructing justice on a global scale. After his promotion to cardinal, he was put in charge of the so-called "Congregation for the Doctrine of the Faith" (formerly known as the Inquisition). In 2001, Pope John Paul II placed this department in charge of the investigation of child rape and torture by Catholic priests. In May of that year, Ratzinger issued a confidential letter to every bishop. In it, he reminded them of the extreme gravity of a certain crime. But that crime was the reporting of the rape and torture. The accusations, intoned Ratzinger, were only treatable within the church's own exclusive jurisdiction. Any sharing of the evidence with legal authorities or the press was utterly forbidden. Charges were to be investigated "in the most secretive way ... restrained by a perpetual silence ... and everyone ... is to observe the strictest secret which is commonly regarded as a secret of the Holy Office ... under the penalty of excommunication." (My italics). Nobody has yet been excommunicated for the rape and torture of children, but exposing the offense could get you into serious trouble. And this is the church that warns us against moral relativism! (See, for more on this appalling document, two reports in the London Observer of April 24, 2005, by Jamie Doward.) Not content with shielding its own priests from the law, Ratzinger's office even wrote its own private statute of limitations. The church's jurisdiction, claimed Ratzinger, "begins to run from the day when the minor has completed the 18th year of age" and then lasts for 10 more years. Daniel Shea, the attorney for two victims who sued Ratzinger and a church in Texas, correctly describes that latter stipulation as an obstruction of justice. "You can't investigate a case if you never find out about it. If you can manage to keep it secret for 18 years plus 10, the priest will get away with it." The next item on this grisly docket will be the revival of the long-standing allegations against the Rev. Marcial Maciel, founder of the ultra-reactionary Legion of Christ, in which sexual assault seems to have been almost part of the liturgy. Senior ex-members of this secretive order found their complaints ignored and overridden by Ratzinger during the 1990s, if only because Father Maciel had been praised by the then-Pope John Paul II as an "efficacious guide to youth." And now behold the harvest of this long campaign of obfuscation. The Roman Catholic Church is headed by a mediocre Bavarian bureaucrat once tasked with the concealment of the foulest iniquity, whose ineptitude in that job now shows him to us as a man personally and professionally responsible for enabling a filthy wave of crime. Ratzinger himself may be banal, but his whole career has the stench of evil—a clinging and systematic evil that is beyond the power of exorcism to dispel. What is needed is not medieval incantation but the application of justice—and speedily at that".

The church still covers up the pedophilic acts against children and as the article above stated is practicing criminal activities. How do you call this man the Pope, a Holy Father when he covers up the rape of children all over the world? For the first time in 600 years this Pope has to step down under new allegations of child rape happening inside the church that he covers up. So I am bringing this up to show you this point. All the pedophilic crimes committed by priests inside the church have not been enough to deem it dangerous. It would be very accurate to label the Catholic Church as a Children Threat Group (CTG) but they won't do it. However they think they can condemn the NOGE by calling us a STG and a gang. More importantly the NOGE is being denied legitimacy as a God Centered Cultural Path to God, by people who belong to a religion that allows these kind of crimes. How can they call my Father out of his name or defame the character of the Nation he built with all the wickedness they have practiced for centuries? They have no moral authority and I give them nothing but crushing truth to destroy these liars and their lies. We don't have to make up names to call them their evil and wickedness is very real. The NOGE is faced with a multi-pronged effort to make us something we are not. They are using government agencies like DOC's, courts, and law enforcement in *combination with the church to discredit our truth of the Blackman being God. That's what this is really all about. They want to stop us from teaching and keep the 85% from learning about the Blackman God of the Universe. History shows they start by talking, then move to talking bad about a people they want to target. Then they send their social scientist to come amongst us and do trading. Those who come amongst us as traders trade in lies and deceit and leave amongst us traitors who will aid and abet them in the accomplishment of their ultimate goal. Once a majority of public opinion has been won by the liars, the process of eliminating the target population begins. Who cares about what happens to a Gang or a dangerous Security Threat Group? The answer is nobody. So we have to stop them from making those lie filled names stick to us. We must stand up and stop their process where it is right now. We cannot afford to lose the war of words especially when the words we speak are the truth. Allah never did and the NOGE will not fail to defend our Name, our God Centered Culture or our future. We will save ourselves and our babies from those who target us for destruction. We are Allah's God's and this Nation united can never be defeated! As I continue to honor Allah I accept the responsibility as a Cultural warrior of the NOGE to do the work that is needed to free our people from Theological discrimination. Until next we build I remain.......... Your True Brother In The Struggle Born King Allah *7)*

CHAPTER 18

ATTEN-TION!

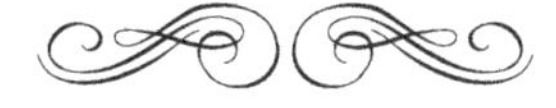

"I was in the Army. I don't teach them not to go in the Army. I went and I came back. I saw action. I got the Bronze Star. I got all these medals. Now I came back home. I didn't benefit from it as by such as luxury but I did benefit by knowing how to teach a man. To teach a man to make a boy a man. This is what I did."

—ALLAH

WHAT IS THE military? In simple terms, the U.S. Armed Forces are made up of the five armed service branches: Air Force, Army, Coast Guard, Marine Corps, and Navy. There are three general categories of military people: active duty (full-time soldiers and sailors), reserve & guard forces (usually work a civilian job, but can be called to full-time military duty), and veterans and retirees (past members of the military). Allah served in the army from 1952 and was honorably discharged 1958. The army exists to serve the American people, to defend the Nation, to protect vital national interests, and to fulfill national military responsibilities. So we salute our veterans, retirees and service men and women on active duty at home and abroad. We want all our service men and women to know we appreciate their sacrifice and courage.

How many Blacks sacrificed theirselves to protect the liberties of theis country? The first person to be killed in the American Revolutionary War was a Black man named Crispus

Attucks. What about harriet Tubman? She was not only the conductor of the "Underground Railroad" responsible for freeing countless slaves, she also served as a nurse and a spy for the Union Army. She was perhaps the most underrated and underappreciated person of the Civil War period. No people described as African-American should be denied those rights or liberties established by the U.S. Constitution. As we come to learn and know our rights, we want those rights to be respected and protected. These are not necessarily rights a man gives to you. These are inalienable rights you are born with as a human being. As Gods and Earths, we see ourselves as the highest form of life (supreme) living in accordance with the law. Therefore, the supreme law in a civilized society is not in disagreement with our way of life. To learn about these rights, I suggest that we learn about the Universal Declaration of Human Rights, the Bill of Rights, and the U.S. Constitution.

For many military veterans of foreign wars, their tours of duty have ended, but they continue to fight for better treatment right here in America. Allah served in the Army during the Korean War and taught his Five Percenters to respect the government and the American flag. We love America just as any other American because we have fought, bled, died, slaved and contributed so much to the very existence of America. Therefore, we demand no less than all of our rights to be respected and protected. We want our flag to be equally respected as we respect the flag of this country and of other nations. While we aren't advocating for people to join (or not join) the military, there are Gods and Earths (and their children) who've served in the armed forces and are alive today to share their stories. For example, Elevation Allah, has had knowledge of self for twenty eight years. He joined the military in 2009 and shared with me the growth and development of his own thinking. Elevation explained to me that good, bad or indifferent, this is still the greatest country on Earth because of the privileges and rights it affords its citizens compared to other countries. He explained how our ancestors built this country and as his perspective changed he embraced change. After joining the military, he learned levels of discipline and training that many of us should have. We should know how to remain calm under pressure and how to defend ourselves, our families, and our property. We should know how to speak to different people, when to speak and when not to speak. *"If you can't deal with a person screaming at you, how can you deal with someone shooting at you?"* said Elevation about military training. He talked about two types of training, technical and tactical. He explained that technical training teaches you how to perform certain functions and tactical training teaches you how to put those skills and functions to practice. Many people know what the words Loyalty, Duty, Respect, Selfless Service, Honor, Integrity, and Personal Courage mean. But how often do you see someone actually live up to them? Soldiers learn these values in detail during Basic Combat Training (BCT), from then on they live them every day in everything they do — whether they're on the job or not.

A sincere Five Percenter (God or Earth) knows the values of Loyalty, Duty, Respect, Selfless Service, Honor, Integrity, And Personal Courage is taught within the lessons we study — Truth, Freedom, Justice and Equality. We teach who and what we are it by exemplifying these values in our interactions with all people. Our teachings are designed to shape boys into men and girls into women. Furthermore, Allah didn't raise punks, complainers and whiners. His desire was for us to be healthy, strong and good breeders. The determined idea is to go from being the *most low* in attitude, thinking, charcter and morals to the *most high* in attitude, thinking, character and morals, not the other way around. Five Percenters are seen and heard everywhere around the world and one could be sitting or standing next to you without you even knowing.

CHAPTER 19

The Future is Written in the Stars

IN THE BEGINNING of this book, the signs of the stars were used to reference the point in time and the conditions we're living in. Just as astrologers use the signs in the stars to predict the future, we use our stars right here on Earth to predict our future. The emphasis we place on the babies being the greatest should support transformational leadership and educational experiences through which children and adults are inspired and enabled to create positive change within themselves, their families and communities.

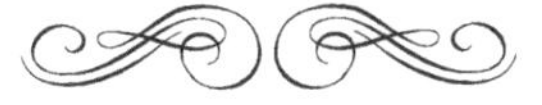

"There is so many Five Percenters it's something else. There still is nothing I can do if you don't show the child in different ways that you care about the child."

—ALLAH

It is our choice to be committed to excellence, integrity, and service. No matter what country, state or city we are in, we can develop and implement:

- Educational opportunities and positive experiences for youth and those who care about them;
- Multicultural and transformational leadership programs which uses art, music, dance, sports, health, wellness, fitness, recreation, environmental education, cultural exchange, and service;

- Learning experiences that promote self-esteem, self-respect, leadership, dignity, and responsibility to be healthy individuals who resolve conflicts in peaceful ways;
- Opportunities to serve, respect, and accept individuals, families, and communities of all cultures, locally and globally.

The future of our nation is written in the stars. The future is written in the very 23 chromosomes each parent passes on to a child. All of our genetic information is written in biological code and passed to our little ones. So our teaching and nurturing of our children is vital to the growth and development of who and what they'll one day become. The beautiful thing about this is we can choose to do something about this today. It's going to take a village to repair many broken families but it's something worth doing.

The babies are a focal point in the advancement of people. Teaching young people is the best method to advance a nation. The allegorical story of Yacub mentioned in our lessons teaches us that the growth and development of children was vital to the making of a new kind of people. In Yacub's process, the birth and survival of a new kind of people was dependent on the destruction of the first (original) people. That process focused on the newborns

to sixteen year olds. In order to make a better world, we have to start with our newborns to sixteen year olds. Each of these stages newborn (ages 0–4 weeks); infant (ages 4 weeks–1 year); toddler (ages 1–3 years); preschooler (ages 4–6 years); school-aged child (ages 6–13 years); and adolescent (ages 13–19); presents new levels of learning for children.

We started as teenagers that mingled with each other as we built. We loved talking about the culture and it wasn't boring. We

SCIENCE FAIR

kept it alive and passed it to our peers, therefore, every Parliament and rally in every region on the planet should have a coordinator who organizes children's events where the youth can meet each other, mingle, and build. Allah intended for us to build a nation or be destroyed. He aptly said, ***"if you stop them [the youth] from mingling together it's over with. It's over."*** [***emphasis mine***].

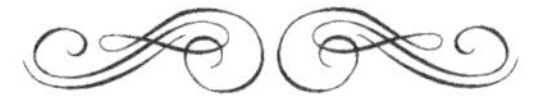

> **"The wealth of a country is the children. Not the money. Once upon a time there was no money. The wealth of any country is the children. And if you don't keep the young people strong, how you going to win?"**
>
> —ALLAH

How many parents lie to their children instead of teaching them the truth head on? Lying to children cripples them, but being truthful empowers them. Covert marketing, misleading advertising, and false claims miseducates parents and children at large, saying only what is necessary to gain sales. Teaching young people to complete school, get a degree, a trade, a skill can save their lives and help them become a responsible adult which is needed for the development of a nation.

If we want to teach our children how important education is, we have to be examples of that. If we want to teach our children righteousness, we have to be example of that. If we want to teach our children to be the makers and owners of tomorrow, we have to prepare them for that by being examples and good teachers today. Remember, actions speak louder than words, and children learn from what we do moreso than what we say.

In this nation, our first form of education is Supreme Mathematics. We start to learn about the quantitative and qualitative values of numbers in relation to our Supreme Alphabets as they are applied to ourselves and everything in the universe. But what happens next is up to the parents. Some homeschool while others send their little ones off to public, charter or private school. Homeschooling provides the most focus on your child because YOU are the teacher and YOU control the curriculum. On the other hand, the moment your child is sent off to the care and custody of an institution to learn, YOU place your child's mind in the hands of an institution. Many active parents know this so when their children come, the curriculum and homework is looked at. Some parents go steps further and teach their children additional things at home while the child is in school. Statistics show more children who get taught at home while in school tend to excel further than those who don't. According to home-school.com:

- 73% of parents who teach at home do so because of dissatisfaction with the American school system.
- Over 2.04 million students are now learning at home, a 75% increase since 1999.
- In the subjects of reading, language, math, science, and social studies, homeschooled children out-perform public school children.

- When it comes to communication, daily living skills, socialization, and maturity, homeschoolers outscored public school children on every level.*

America is the most powerful country in the world, yet lacking behind other countries in education. More importantly, education statrts at home, not in any other institution. The home is where children are first taught by their parents, their values, their morals and their outlook on the world. As Five Percenters, we are concerned about our children being properly educated. SciHonor Devotion is one of many sisters in the nation who took matters into her own hands to educate her children. Here, she offers a comprehensive checklist of lessons any concerned parent can teach their children:

26 Lessons to Teach the Babies by Sci-Honor Devotion

1. Let us teach our babies from the time of conceptual thought who they are and their origin in this world and beyond.

2. Give them mighty names and the knowledge of their names so that they begin to build great character and self-esteem.

3. Teach them their strong legacy and all of the power that they hold.

* Source: http://www.home-school.com/news/homeschool-vs-public-school.php

4. Teach them pride and humility so that they will desire to constantly learn.

5. Teach them the value of their mental and physical temples and how to care for them, protect them and how to eat to live.

6. Teach them how to have respect for themselves and for others.

7. Teach them about honesty and truth, so that they will know that the truth will always shine through.

8. Teach them gratitude and appreciation, so that they never take the jewels of life for granted.

9. Teach them discipline and how to be responsible for their ways, words and actions, while being prepared for whatever justice (reward or penalty) comes their way.

10. Teach them basic manners and how to be good communicators so that they may be able to solve conflict without killing each other.

11. Teach them integrity and how to stand firm on their values (squares) setting standards and living up to those standards.

12. Teach them the value of honesty and the value in keeping their word.

13. Show them how to be diligent and how to endure while striving for excellence, maximizing on their power and not settling for less. Perseverance is to be honored.

14. Teach them what "true" peace is and its value behind this verbal greeting.

15. Guide them to freedom, so that they know what they are striving for.

16. Teach them discernment and dissection so that seeing things clearly will come to them.

17. Show them why knowledge is the key and the foundation.

18. Teach them self-control and how to handle their emotions.

19. Teach them the science of balance and how to maintain when walking the line and knowing the ledge.

20. Show them why knowledge is the "key" and the foundation to all things.

21. Teach them their "infiniteness", "eternalness", and "foreverness".

22. Teach our sons what their roles will be when they are the kings of their families and how to be fathers.

23. Teach our daughters what their roles will be when they are the Queens of their families and how to be mothers. These skills are best taught when watching us, the parents in these roles, fulfilling and mastering our duties.

24. Provide them with love, peace and happiness.

25. Teach them who God is so that they waste no time searching.

26. Lastly, but not least, give them the mathematical keys to the universe so that they may enter any door when we aren't there to guide them.

CHAPTER 20

Power (5) of the Cipher (0)

FROM A SCIENTIFIC and mathematical perspective, POWER is the rate of doing work. People have the capacity to do great work. As civilized people, we have knowledge, wisdom, understanding, culture, and refinement. So, we have Power or Refinement working cohesively to make a person, place or thing better or improved over time based on the amount of work that takes place over time inside and outside that cipher *(you).*

In our Supreme Mathematics, (5) is the principle of POWER. The expression Five Percent is five out of one hundred. Zero (0) is the Cipher. So, five people out of every one hundred will have the power to teach and work to make the state or condition of a cipher (person, place or thing) better or improved. I used to think everyone in the world would eventually come to their senses, unite and work together to make things better. Since all people are not like-minded and are divided so easily, only the few who understand will do the work that must be done. We all may not share the same views, however, we are the same people in a similar condition with similar problems. This unites us all in one reality.

Regardless of how many ciphers there are, there is only one cipher which is true in mathematics as well. Therefore, we have to learn to use our power when we find that power within ourselves. Trust me, it is there. Improvement comes about through the work we do over time inside oursleves and outside ourselves. The work that is required of us all will bring about the change we seek in ourselves and around us. *Power Cipher* (or 50) is *power to the people, power in the people, and power for the people* no matter where they are in life or what they've been through because a cipher is any person, place, or thing. In people, places and things there is surely power. These teachings are about showing you where and how to get that great power and put it to good use. That's cause for Jubilee. In Judaism and Christianity, the concept of the Jubilee is a special year of remission of sins and universal pardon. In the Biblical Book of Leviticus, a Jubilee year is mentioned to occur every fiftieth year, in which slaves and prisoners would be freed, debts would be forgiven and the mercy of God would be particularly manifest. Fifty years can mean different things to different people, however, our culture has been challenged, tested, and have gone through the fire. We withstood the test of time and we are still here teaching on.

Don't wait until tomorrow, next week, next month, or next year to tap into your power. Empower yourself and you will inspire and empower others. You can empower your WILL to do it, but only if that's what *you* want! The speed and power of a thought is based on the sincerity of that thought. Let's energize and empower each other by coming together and making things happen. As our power goes from knowledge to born, remember that the power of the cipher is the power of the people in that cipher!

ALLAH'S WORLD MANIFEST

The First Universal Parliament in April 1967.
Courtesy of C Allah, Fort Greene, Brooklyn, (the 'Head of Medina').

JUSTICE WITH YOUNG GODS

LANERS

A 'NATION' UNTO THEMSELVES

Courtesy of Peace TySun Allah

Courtesy of Lord Dumar Allah

Universal Parliament in Mecca, NYC, Late 60s/Early 70s

L TO R: Wisdom God, Truth Allah (Izame), Understanding Seed Kamile Allah, Knowledge Allah (Dihoo), Tyshon Allah, King Wisdom Allah and Blackseed Jarmean Allah; 1970s, Asbury Park, New Jersey (Now Justice); Courtesy of Rahe Allah

Queens at 1971 Show & Prove

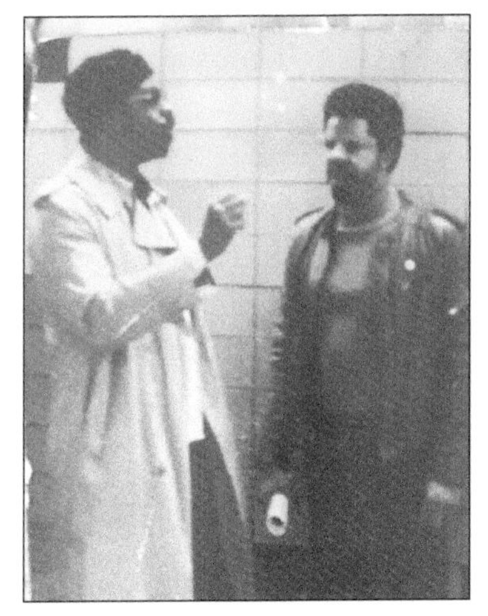

Universal Shaamgaudd Allah builds at the Universal Parliament, Early 1980s.

Dumar Wa'de Allah builds at NYC Universal Parliament, 1983. Courtesy of Peace TySun Allah

Shaqun Asia, Heart of Medina, 1984. Courtesy of Jamel Shabazz

Flatbush 1984. Courtesy of Jamel Shabazz.

The Gods on 125th Street, 1988. Courtesy of Jamel Shabazz.

Earths in Fort Greene Rally (Medina)

"JAMEL'S WORK IS LIKE THAT. HIS WORK IS THAT VOICE THAT WAKES YOU UP GENTLY, SUDDENLY, AND LIKE ANY GREAT ARTIST MAKES
YOU SEE THINGS WITH FRESH EYES. JAMEL'S ART IS A BLACK MIRROR, REFLECTING WHERE WE'VE BEEN, WHERE WE ARE, AND WHERE
WE MAY BE HEADED.... HE USES HIS CAMERA LIKE KRS-1, CHUCK D, OR GIL SCOTT-HERON USE THEIR MICROPHONES, TO REFLECT A
CULTURE, AND THE HOPES, DREAMS, NIGHTMARES, AND HISTORIES OF A PEOPLE."
-ERNIE PANICCIOLI

Kings Park in the Oasis

Courtesy of God Barsha Allah

Gods in the Head of Medina (Fort Greene)

Gods and babies from Allah's Paradise (Asbury Park, NJ) Feb. 22, 2015

Stars at the Show & Prove

2122

2015 Earth Appreciation Day, Afrikan Poetry Theatre (Oasis)

GIRLS

IN LOVING
MEMORY OF

Medina (Brooklyn, NY)

The Oasis, Allah's Projects (Astoria)
Courtesy of Allah Mathematics

Gods in the Desert/Oasis

Aids Walk in Central Park 2014

The Pelan Council

ABG #7, Gykee Mathematics and Dumar Wa'de Allah in Mt. Morris Park (aka Marcus Garvey Park), Mecca

PRESIDENT
BARACK OBAMA

2122
#XOX

SAVING THE CHILDREN
THE BABIES ARE
THE GREATEST
PELAN COUNCIL

CLOCKWISE FROM TOP: Queen TruEarth, Original Love Reflection, I-Queen Divinity, Lovasia Sunblessed Graceful Earth, Queen SheReal Earth, Earthly Paradise, Queen Sheraina Earth, Queen Omala, Queen Cipher. **BOTTOM:** Beautiful UmiEarth and Queen-Math3matiqs

Gods on the Square, 42nd Street Times Square, NYC.
Photography by King of the Flix aka King Sincere

Savior's Island (Staten Island, NY) Courtesy of Black Cream Allah

Divine Cee (Washington, D.C.), Courtesy of Bar-King Shamell, Samiya Radiant Earth & I Add Uneek

Universal Street Academy in Power Hill (Philadelphia, PA)

Philly Family Day

Lynn (Love Why) & Boston (Black Medina), Mass.
Courtesy of Kingasiatic Allah

Courtesy of Allah Mathematics, BarberBoyz Club

Gods & Earths from Allah's True Land at the 43rd Annual Show & Prove
Courtesy of IAtomic Allah

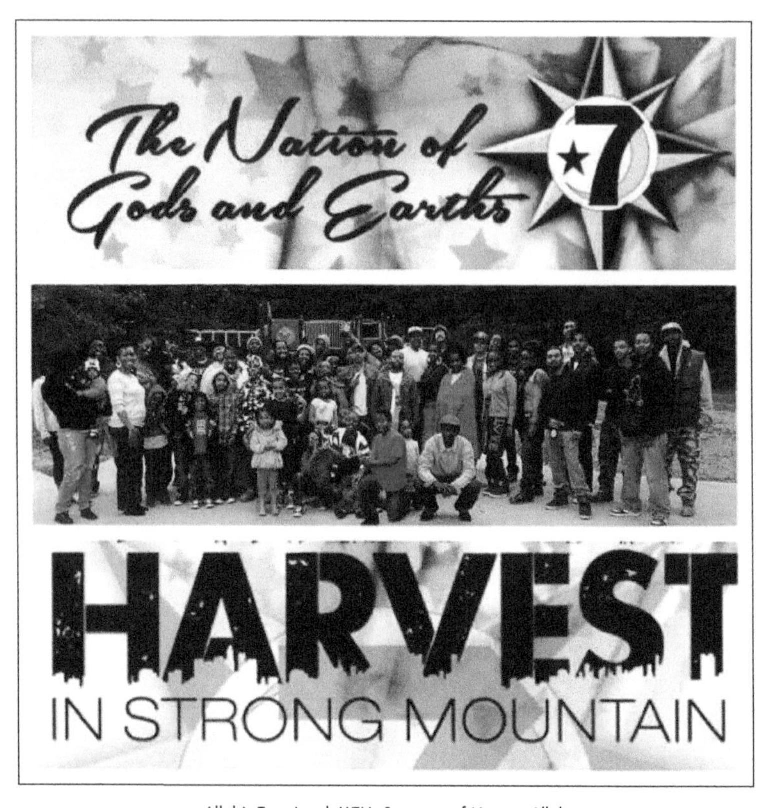

Allah's True Land (ATL) Courtesy of IAtomic Allah

2015 Martin Luther King Day Parade in Allah's True Land (ATL)
Courtesy of Powerful Allah

Victory Allah, Virginia, Courtesy of Earthly Jewels via Jahshamel Ph.d Bishan Allah

Now Cee (North Carolina)

Gods, Earths and babies in Now Cee (North Carolina)

Family Reunion Region 4, Courtesy of Almighty Unikue U Allah

2012 Show & Prove, Now Cee, Courtesy of Omni Photography

Savior Cee (South Carolina), Courtesy of Infinite Allah aka Doc Divine

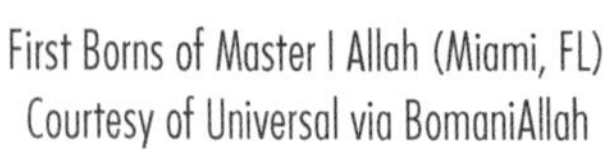

First Borns of Master I Allah (Miami, FL)
Courtesy of Universal via BomaniAllah

The Sudan: Courtesy of Lord Hashim

Photography by Jirard

D-Mecca (Detroit, MI)

The Saudi Cipher (St. Louis, MO), Courtesy of Hakim IsAllah & Power God Allah

Cream City courtesy of SelfKingdom IsAllah

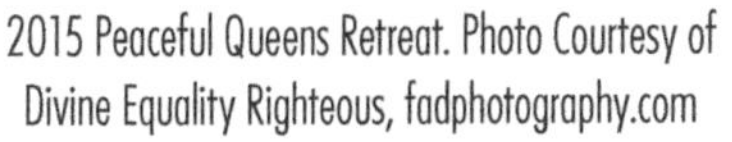

2015 Peaceful Queens Retreat. Photo Courtesy of Divine Equality Righteous, fadphotography.com

The Universal Family in part of Region 6 Southern Ontario/Western New York; Truth Cipher [Toronto Ontario, Canada], Steel City [Hamilton Ontario, Canada], Atlantis [Niagara Falls, NY] and Bethlehem [Buffalo, NY]. Courtesy of Saladin Quanaah Allah

(LEFT) MCs Kasim Allah, Rashad Sun and Lakim Shabazz. (ABOVE) Khalik Allah and Jamel Shabazz by Ryan Lyons Photography

LOVE
YOUR
PEOPLE
every
DAY

LOVE
YOUR
PEOPLE
DAY

THE 50TH ANNIVERSARY CONCERT AT THE WORLD FAMOUS APOLLO THEATRE
HARLEM, NEW YORK 2014

Sophisticated Gentleman

Photography is the science, art and practice of creating durable images by recording light or electromagnetic energy.

Photos by Khalik Allah

EXIT

NY
1915

Allah
Nation Builder
Region One
'The Foundation'

PEACE

Appendix

UNIFICATION IS THE DIRECTIVE, TRUTH IS THE CRITERIA

Peace to the nation.

Elamjad here. This issue's editorial is for scientists. It presents some complex concepts that the average nation member may have to chew on a bit before digestion. Mainly [...]an address to the nation's social architects and its leadership. So if some of its escapes the reader I admit that it is mainly geared for those that it does not, so bear with me.

They are beginning to refer to it as the era of accountability. The term suggests that we have arrived at a historic impasse in the nation's history. It suggests that we are entering an era in which what is called "the nation," that largely undefined congregation bound more by ideology, than material structure, is becoming a more mature body-politic. It seems we are growing less willing to accept guidance by circumstance. The consensus is seemingly more and more leaning towards pro-active measure. Our condition, it is argued, begs more standardizations. More predetermined processes and procedures. We are growing increasingly unwilling to allow things to simply peter out....so long as it doesn't cost us. Accountability applies to a lot more than just misdeeds.

There is clearly a work in progress here. Gods and Earths nationwide are implementing different initiatives. Some of these initiatives are complex tasks with vast challenges to surmount. The National Treasury initiative implemented by the AUD is one such example. -[...] How will they achieve widespread conformity to the plan? The formulae for solving this problem are surely elusive, but not non-existent. The money by and large universally proves to be the great barrier between the committed and the non-committed, between mobility and stagnation. With us the nation, maturity at this age means that we understand building without falling victim to the devil's civilization. So emphasis on material is suitable for the time. Knowledge is forever our foundation, and the achievement of individual growth based on our teachings is always prerequisite. For man though, particularly those long tenured and well grounded, the issue of material commitment is more pronounced. After many years in the nation it is expected that one's foundation is secure. Such emphasis on material showing and proving doesn't pose the same potential hazard to self-development as may affect one still finding him or herself.

Lately I've heard a recurring theme espoused by certain national officials. Financial regulation and taxation. The theme advocates a levy upon nation related commerce and eventually on the private incomes of individual nation citizens. *"We are not going to let just anyone sell stuff with the flag on it,"* a repeated declaration. *"Eventually you are going to have to pay taxes,"* is another assertion in such speeches. In theory the idea of imposing a fee on anyone using or selling items bearing the national standard may have merits worthy of exploration, but there are a plethora of issues and associated minutia which any plan deemed "national" must speak to. There are many prerequisites which must be satisfied for even the slightest chance at success. The integrity, qualifications and track records of the plan's proponents are not the least of these. Whenever there is advocacy on behalf of any so-called "nationwide" application of rules, especially for some plan that places material demands on nation members, those behind the plan had better demonstrate their selves to be highly competent. Often, short sightedness based on lack of serious assessment of the social dynamic on a vestigial level leaves some sorely qualified to bring the nation usable solutions. The issues of identification on a national scale must receive clarity before any insistence on "nationwide" obligation can be applied with supreme equality and justice. I suspect we are a very long way off. Consider following.

Having for many years evaluated and experimented with this complex dynamic of the mandate to build our "nation," I find that the more advances are made in earnest, the clearer the presiding complexity of this task becomes. In this regard, I would further add that my reference seeks to elaborate prerequisite to any bona fide attempts at building materially on a (and I use the term loosely) "national" scale. In fact, much of our "nationhood, with respect to any meaningful efforts to achieve economic buoyancy, if not sovereignty, demands clear resolution on the essentials of unification. This same reasoning also applies to other forms of social construction besides economic development.

Our *"nation"*. In a material sense, what is that? Who is that? Meaning who are its members? What is the census? What are the criteria for identifying them to the extent of good standing?

That is their mandatory obligation to the collective. What is their economic duty, their military obligation to service? What is their function in a decision-making capacity and or hierarchy? What are the shared resources of the nation, save Allah School, if that? Who says who is, and who isn't? Also, what is the basis and means of enforcement? Ultimately who specifically, and in what numbers, shall be obligated to, and inure to the benefits of such efforts once accomplished? Moreover how are these processes achieved, over what time period? Is there a progressive evolution perhaps, or some other installation or infusion of the requisite systems? Absent of the tangible usability of answers to such queries, we shall be loath to proceed successfully. Furthermore, this reality applies to any number of projects whose scope and magnitude encompasses the entire "nation." This is serious business, perhaps the most serious in uncharted regions. It is only my intention to elucidate, even in some small sense, some of the preliminary work required and challenges to surmount as one would enjoin this magnanimous nation-building endeavor. It is not my purpose to cast doubt, but rather to indicate the importance of thorough assessment and the consequent preparations. These are absolute imperatives if the matter before us is to be given any serious attention.

For so long the superficial views taken of our social configuration offers no real detailed answers to creating the needed infrastructure that yield usable results for the nation. Can we even define a "nation member" to the extent that anyone proposes to hold "all nation members" to task for anything, let alone a financial obligation? To this extent of accountability, exactly who is "a nation member?" Certainly to achieve acquiescence to a national agenda those responsible for adhering to it must be defined.

There are alternatives however. National plans of present employ the principle of voluntary compliance of the willing. But what shall compel us to act in concert, to move a nation in lockstep? What shall persuade us en masse to pledge our goods and our persons to achieve what we want? Is there another way? Do we await a mysterious benefactor? To aspire towards the hope for donations as a first line strategy is to contend that someone else should pay, and not you. To wait for others to perform as a primary option is irresponsible. We must each feel personally responsible, especially the leadership.

Paying into the nation's newborn financial system is the litmus of sincerity. I suggest those advocating accountability, not merely for misdeeds, but for financial and economic responsibility first take stock of their own performance. Are you meeting your National treasury obligation? Do you pay into the system? The leaders will show and prove their right to appropriate authority by complying now absent of a mandatory system. For even absent of such a system the mandate must be made real by the voluntary compliance of the front-liners. For sincere leaders the mandate already exists and will be evidenced in each one of them by their act of doing now what is proposed each of us will have to eventually do by law. Who shall lead? Will it be you, and will it be enough? If you're not paying into the system, you need to stay the hell out of my face. How can you demand of others what you will not lead the way in yourself? (Elamjad Born Allah, '*The NGE Power*', *Issue #1.4, April 2003, p. 2*)

The P.E.A.C.E. Course

Political education and civilization enrichment are valuable and necessary components in any educational curriculum especially within this culture. In the urban community, there are countless young people with an abundance of energy and enthusiasm to be leaders. There is also a burning desire in young people to want knowledge of self and understand the world they live in. Political education is needed for any person engaged in using wisdom and diplomacy to shape policies and government. Political education is integral in understanding and respecting the rules of our culture and other cultures. Civilization enrichment is therefore a necessary component for cultural development. If a person wants learn how to navigate and apply the lessons we study, increase the value or quality of their life or the lives of others, or learning methods of

teaching civilization to those who are uncivilized, they can come to the P.E.A.C.E. course at Allah School in Mecca. The course is instructed by Sunez Allah and people are able to discuss a wide range of topics that delve into our teachings, including finding and maintaining Peace in themselves in order to teach others.

The P.E.A.C.E. (Political Education And Civilization Enrichment) Course is taught at Allah School In Mecca (2122 7th Ave., NY, NY 10027) every Thursday night from 7 PM to 10 PM. It is a Civilization Class sharing our teachings (Knowledge of Self through Supreme Mathematics along with Supreme Alphabet and 120 Lessons) through a complete curriculum of lifestyle training including health and training protocols, reading, writing and communication skills work instructed by Sunez Allah. Lessons, books, assignments, resources and references are included FREE to students. For more information contact: sunez97@gmail.com.

Acknowledgments & Credits

Master King Justice-U-Allah Forever, Sunez Allah, Shahid M. Allah, Jamel Shabazz, Khalik Allah, Kamell Allah, Life Allah (CA), Original Author Allah, God Supreme (Mecca), Cee Allah (OGDC Chairman), Allah Divine, Born King Allah (NC), Peace TySun Allah, Supreme Understanding Allah, Tau Justice, Shabue Allah, C Allah (Fort Greene), Gykee Mathematics, ABG, Infinite Al-Jamaar-U-Allah, Um Allah, Lord Jamar (Brand Nubian), God Emblem, Lord Hashim, Allah Intelligence, Bo'kem Allah, Emblem, Lord Jamar Allah, Elevation Allah and Lord Hashim Allah. DJ Wise (aka Akeem Rashad Allah), Sha-Born Intelligence, Life Supreme Allah, God Universal, Rondu Allah, Supreme (First Choice Apparel), Great God, Powerful Tayshawn God Allah, Knowledge Scientific God Allah, Raking Allah, Lord Rondu, Raliek Allah, Knowledge Born Allah, Young Divine, Unique, Reality, Ka-shawn, Ruler Divine, King Sincere, Laking, Prince Love Allah, Cream Allah, and Understanding Scientific Allah (PBUH). Peace to all the Gods! Lovasia Sunblessed Earth, Victorious Lanasia Earth, Sci-Honor Devotion, Earth I Asia, Queen TruEarth, Earth Izayaa Allat, Equality Perfection, UmiSudan Lunar Allat

Wisdom, E-Queen Glorious Earth, Original Love Reflection, Earthly Paradise and Earthly Embrace. Peace to all the Earths and Queens! Peace to all who celebrated our 50th Anniversary with us at the Apollo including but certainly not limited to: the World's Famous Supreme Team, Erykah Badu, Big Daddy Kane, Freedom Allah [Poppa Wu], Brand Nubian, Kasim Allah, Capone-N-Noreage, Tragedy Khadafi, Royal Flush, Roxanne Shante, Hakim (of Channel Live), DJ Kay Slay, DJ Wise, DJ Imperial, Black Page and Brown Mood Consulting, Papoose, Remy Ma, Frukwon, Rashad Sun, Sha Rock, Narubi Selah, Sophisticated Gentlemen, A-Alikes, Force M.D.'s, Spinderella (Salt-N-Peppa), Busta Rhymes, Smif-N-Wesson, InfMega Media; all those who submitted photos. Peace to all the young Gods, Earths and Queens! PEACE!

> *"It's our duty to each other to teach and exemplify respect, love, peace, happiness, and unity on the basis of our historical and national nobility. Furthermore, it's our destiny and duty to our future to call ourselves to order and build regardless of name, nationality, title or geography."*
>
> – Starmel Allah

To order part 1 of The Righteous Way, a set of both books, wholesale orders, view the latest articles, photos and videos, visit us at:

WWW.THERIGHTEOUSWAYBOOK.COM

CPSIA information can be obtained
at www.ICGtesting.com
Printed in the USA
FSOW03n1350210316
18246FS